I0814687

# PATENTING LIFE

# PATENTING LIFE

## *Tales from the Front Lines of* INTELLECTUAL PROPERTY *and* THE NEW BIOLOGY

JORGE GOLDSTEIN

Georgetown University Press / Washington, DC

Library of Congress Cataloging-in-Publication Data

Names: Goldstein, Jorge A., author.
Title: Patenting life : tales from the front lines of intellectual property and the new biology / Jorge Goldstein.
Description: Washington, DC : Georgetown University Press, [2025] | Includes bibliographical references and index.
Identifiers: LCCN 2024018103 (print) | LCCN 2024018104 (ebook) | ISBN 9781647125196 (hardcover) | ISBN 9781647125202 (ebook)
Subjects: LCSH: Biotechnology—Law and legislation—United States. | Patent laws and legislation—United States. | Intellectual property—United States.
Classification: LCC KF3133.B56 G647 2025 (print) | LCC KF3133.B56 (ebook) | DDC 346.7304/86—dc23/eng/20240528
LC record available at https://lccn.loc.gov/2024018103
LC ebook record available at https://lccn.loc.gov/2024018104

∞ This paper meets the requirements of ANSI/NISO Z39.48-1992 (Permanence of Paper).

26 25 9 8 7 6 5 4 3 2 First printing

Printed in the United States of America

Cover design by Erin Kirk
Interior design by Paul Hotvedt

Disclaimer: The views and opinions expressed here are solely those of the author and are not to be attributed to his law firm or his firm's clients.

To Sandy, my dear wife and best friend

And to the memory of Professor Jack Wands,
long a client and a colleague, who sadly passed away before
seeing the publication of this book.

All creation is a mine, and every man, a miner.

*Abraham Lincoln, 1858*

# Contents

# Acknowledgments

I am deeply grateful to the protagonists whose stories I tell and who gave of their valuable time to talk to me and read what I wrote about them: Niels Reimers, Eldora Ellison, Theodore Friedmann, the late Jack Wands, Stephen Hoffman, David Brook, Philip Felgner, David Felson, Marc Grodman, Elizabeth Holtzman, Loretta Sue Loesch-Fries, Lucas Tyree, Nicolas Cock Duque, Federico Trucco, and Raquel Chan. I learned from them that creativity takes many forms; yet I saw that they all share a single, focused passion for whatever they have given the world, whether in medicine, science, law, or business.

A personal thank you to my dear friend Marvin Guthrie, who at different times in our professional lives was a law partner, a client, and always a wise collaborator. I learned from him how to practice law in an ethical and human way. I am indebted to several of my law partners and professional colleagues with whom I discussed plenty of fine legal points: Tim Shea, Eldora Ellison, and David Cornwell. Thanks to Mike Ray, my managing partner, who graciously looked the other way when my billable hours became consumed by writing this book.

I thank Larry Staples, my Jungian friend, who taught me how to find creativity in the discord of opposite views. I found inspiration, camaraderie, and advice in the members of my Wednesday writing circle: Alan Miller, Laurence Carter, John Hasse, and Peter Eisner. They patiently read and thoughtfully critiqued the chapters as I was creating them. They helped turn the wooden prose I had acquired, first as a scientist and then as a lawyer, to be a touch more literary. And a thank you to our circle's writing coach during many years, Gina Hagler, who provided many lessons in good writing.

Several colleagues, friends, and acquaintances—Carla Kim, Gaby Longsworth, Marc Hertzman, Rebecca Hertzman, Miguel Tejblum, Alan

Grodzinsky, and Dan Guttman—read all or parts of the manuscript with curiosity and gave me excellent ideas. My wife, Sandy, and my daughter Thalia also read it, adding love to interest. All of them offered invaluable critiques and gave me suggestions on how to make law and science approachable to less specialized readers.

I am grateful for the feedback provided by my anonymous reviewers. They caught several instances of unclear thinking or expression. The final product is much better because of their careful evaluations. My agent, Jane Dystel at Dystel Goderich and Bourret LLC, and my acquisitions editor at Georgetown University Press, Hilary Claggett, introduced me to the—until then barely known to me—world of book publishing. A big thank you to Sonya Manes, my careful editor at Georgetown University Press, who greatly improved my syntax and clarity, and to Elizabeth Sheridan-Drake, who kept everything running smoothly.

All remaining errors or omissions are, of course, mine alone.

# Dramatis Personae

| | |
|---|---|
| Allison, James | One of the scientists who developed the theory of immune checkpoints; Nobel Prize winner (2018). |
| Ausubel, Fred | Professor (1981), Department of Molecular Biology at Massachusetts General Hospital (MGH). |
| Barañao, Lino | Argentine minister of science and technology (2007–19) and co-inventor of transgenic cows that produce human growth hormone in their milk. |
| Beachy, Robert | Plant biologist who in 1992, as a professor at Washington University St. Louis was involved in a patent interference on viral protection in plants. |
| Berg, Paul | Professor at Stanford University who in 1971 created the first recombinant molecules using viruses; he won the Nobel Prize in chemistry in 1980. |
| Beutler, Bruce | Professor at the University of Texas whose United States patent application was awarded priority in the interference over the drug Enbrel. |
| Bottazzi, Maria Elena | Scientist at Baylor University, Texas, who in 2022 with her collaborator Dr. Peter Hotez announced the availability of Corbevax, which they called a "patent-free" COVID vaccine. |
| Boyer, Herbert | Professor at the University of California who co-invented recombinant DNA technology and later founded Genentech. |
| Brook, David | Attorney for Centocor, a licensee of the Massachusetts General Hospital during the *In re Wands* case (1988). |
| Burger, Warren | Chief Justice of the United States Supreme Court, who in 1980 wrote the decision in *Diamond v. Chakrabarty*. |

| | |
|---|---|
| Chakrabarty, Ananda | Scientist at General Electric who invented and patented microbes capable of degrading oil mixtures; his patent was upheld by the Supreme Court in 1980 in *Diamond v. Chakrabarty.* |
| Chan, Raquel Lia | World-renowned expert in photosynthesis and the inventor of multiple drought-resistant seeds, Chan is the director of the Agrobiotechnology Institute of Santa Fe and was named one of the ten most outstanding women scientists in Latin America by the BBC. She did her undergraduate studies at the Hebrew University of Jerusalem and received a PhD degree in biochemistry from the National University of Rosario, Argentina. |
| Charpentier, Emmanuelle | Scientist who, together with Jennifer Doudna, won the Nobel Prize in chemistry in 2020 for their invention of CRISPR. |
| Chilton, Mary-Dell | Plant biologist at Syngenta who in 1983 invented the disarming of the large plasmid of *Agrobacterium* and opened the doors to the genetic engineering of plants. |
| Cochran, Johnnie | Defense lawyer for O. J. Simpson in the 1995 trial of *State of California v. Orenthal James Simpson.* |
| Cock Duque, Nicolas | CEO of Ecoflora Cares, the Medellín, Colombia, company involved in the commercialization of blue color from the *jagua* fruit. |
| Cohen, Stanley | Professor at Stanford University who co-invented recombinant DNA technology and patented it together with Herbert Boyer. |
| Coley, William | American physician who, in the late nineteenth century, observed that bacterial infections can cause tumor regression. |
| Collins, Francis | Scientist at the University of Michigan who, together with Lap-Chee Tsui, first identified and isolated the gene for cystic fibrosis. He was the director of the National Institutes of Health until 2022. |

| | |
|---|---|
| Cook-Deegan, Robert | Principal author of a 2010 study on the gene patent wars surrounding LQT Syndrome, published while he was a professor at Duke University. |
| Crick, Francis | Scientist at Medical Research Council (MRC) in Cambridge, England, who in 1953, together with James Watson and Rosalyn Franklin, elucidated the double helix structure of DNA. |
| De Gama, Mustaqeem | Leader of the South African Permanent Mission to the World Trade Organization who in 2020 helped write a proposal to waive IP rights in COVID-19 vaccines. |
| De Schutter, Olivier | UN rapporteur for food (2008–14) and professor at American University Law School. |
| Doll, John | Director of the Biotechnology Examination Group at the US Patent Office in the late 1980s. |
| Doudna, Jennifer | Scientist who, together with Emmanuelle Charpentier, won the Nobel Prize in chemistry in 2020 for their invention of CRISPR. |
| Ehrlich, Paul | Early twentieth-century German immunologist who conceived of "magic bullets" to treat disease with low side effects. |
| Ellison, Eldora | Law director at Sterne, Kessler, Goldstein & Fox, PLLC; she has a PhD from Cornell University in biochemistry, molecular, and cell biology, and a JD from Georgetown University Law Center. |
| Fauci, Anthony | Former director of the National Institute of Health's National Institute of Allergy and Infectious Diseases who approved a second clinical trial for Sanaria's attenuated sporozoite-based malaria vaccine, this time by intravenous injection. |
| Federico, J. P. | Legal scholar and director of the US Patent Office in the 1940s and 1950s. |
| Felgner, Philip | Scientist who at VICAL, Inc. in the late 1980s, invented the use of genetic vaccines. |

| | |
|---|---|
| Felson, David | Professor of Medicine at Boston University and worldwide expert in rheumatology. |
| Fleming, Alexander | Microbiologist who in 1928 discovered penicillin. while working in St. Mary's Hospital in London |
| Franklyn, Rosalyn | Scientist at MRC in Cambridge, England, who in 1953, together with James Watson and Francis Crick, elucidated the double helix structure of DNA. |
| Freeman, Gordon | Professor of Medicine at Dana-Farber Cancer Institute and Harvard Medical School who worked on the theory of checkpoint inhibition. |
| Friedmann, Theodore | Emeritus Professor of Pediatrics, University of California, San Diego, Dr. Friedmann is considered one of the pioneers in gene therapy and has won several awards for his work, among them the Award of Merit from NIH in 2003 and the Japan Prize in 2015. |
| Fromkin, Drew | CEO (2007) of Clinical Diagnostics, Inc., a clinical diagnostics company involved in a patent dispute on LQT Syndrome. |
| Golde, David | Physician at UCLA Medical Center who in 1976 removed John Moore's spleen and, with Shirley Quan, invented the Mo cell line. |
| Goodman, Howard | Head (1981), Department of Molecular Biology at MGH. |
| Grodman, Marc | World-renowned expert on genetic diagnostics and personalized medicine who in 2007–2008, as the CEO of Bio Reference Labs and its affiliate GeneDx, cracked the thicket of multiple patents that protected the genetic diagnostic assays for the Long QT Syndrome. |
| Gusella, James | Scientist at MGH who first identified and isolated the gene for Huntington's Disease. |
| Guthrie, Marvin | Head of the Office of Technology Transfer of MGH in the 1980s. |

| | |
|---|---|
| Haley, James F. | Scientist-lawyer who received a PhD in organic chemistry from Brandeis University and a law degree from Suffolk University Law School; he is now a partner at Haley Giuliano in New York. |
| Hoffman, Stephen | CEO of Sanaria, Inc., and one of the world's foremost experts on malaria who has developed, by using attenuated sporozoites, one of the most successful vaccines against the disease. While not yet approved, at the time of writing the vaccine is undergoing clinical trials. |
| Holtzman, Elizabeth | Member, House of Representatives from New York (1973–81); as a private lawyer she was involved in the House Hearings on genetic diagnostics (2007). |
| Honjo, Tasuku | Physician-immunologist at Kyoto University who worked on the theory of checkpoint inhibition; Nobel Prize winner (2018). |
| Hotez, Peter | Professor at Baylor University, Texas, who in 2022 with his collaborator Dr. Maria Elena Bottazzi announced the availability of Corbevax, which they called a "patent-free" COVID vaccine. |
| Itakura, Keiichi | Chemist at the City of Hope Hospital in Los Angeles who, together with Arthur Riggs, first produced insulin by recombinant DNA methods. |
| Keating, Mark | Professor at the University of Utah who in the 1990s did groundbreaking research on inherited LQT Syndrome. |
| Köhler, George | Immunologist at MRC who in 1978, together with César Milstein, invented methods of making monoclonal antibodies. |
| Lacks, Henrietta | Cancer patient at Johns Hopkins Hospital whose cells were removed in 1951 and resulted in the HeLa cell line. |
| Loesch-Fries, Loretta Sue | Associate professor of botany, plant path botany, and plant pathology at Purdue University College |

| | |
|---|---|
| | of Agriculture; one of the world experts on AMV viruses who in 1992, as a professor at the University of Wisconsin, was involved in a patent interference on viral protection in plants. |
| Lourie, Alan | Judge, US Court of Appeals for the Federal Circuit (1990–present). |
| Marshall, Barry | Australian physician who in 1984 self-experimented with *H. pylori* as a cause of ulcers. |
| Mendel, Gregor | Nineteenth-century monk and researcher considered to be the father of modern genetics. |
| Michel, Paul | Judge, Court of Appeals for the Federal Circuit (1988–2010). |
| Milstein, César | Immunologist at MRC who in 1978, together with Georg Köhler, invented methods of making monoclonal antibodies. |
| Moore, John | Cancer patient at UCLA Medical Center whose spleen was removed in 1976, resulting in the Mo cell line. |
| Mosk, Stanley | Associate justice of the Supreme Court of California who, in 1990, wrote a dissent in the case of *Moore v. U. California*. |
| Moskowitz, Margaret | Patent examiner involved in the examination of the application that led to the precedent *In re Wands* (1988). |
| Newcomer, Clarence | United States District judge of the United States District Court for the Eastern District of Pennsylvania. |
| Newman, Pauline | Judge, Court of Appeals for the Federal Circuit (1984–present). |
| Panelli, Edward | Associate justice, Supreme Court of California who in 1990 wrote the majority opinion in the case of *Moore v. U. California*. |
| Pasteur, Louis | Nineteenth-century microbiologist and chemist who invented methods of brewing beer by using purified fermentation yeasts. |

| | |
|---|---|
| Plager, Sheldon Jay | Judge, Court of Appeals for the Federal Circuit (1989–present). |
| Quan, Shirley | Scientist at UCLA Medical Center who, with Dr. David Golde, invented the Mo cell line. |
| Quarles William, Jr. | Federal District Court judge, District of Maryland, who presided the litigation over *Phaffia* yeast cells in 1997–2003. |
| Rader, Randall | Judge, Court of Appeals for the Federal Circuit (1990–2014). |
| Raubitschek, John | Associate solicitor for the commissioner of patents, who argued the *In re Wands* case before the Court of Appeals for the Federal Circuit in 1988. |
| Reimers, Niels | Member of Stanford's Sponsored Projects Office who, in 1974, filed the basic US patent application on recombinant DNA technology. |
| Rich, Giles | Judge, Court of Customs and Patent Appeals (1956–82) and then of the Court of Appeals for the Federal Circuit (1982–99). |
| Riggs, Arthur | Geneticist at the City of Hope Hospital in Los Angeles who, together with Keiichi Itakura, first produced insulin by recombinant DNA. |
| Rowland, Bertram | Patent attorney for Stanford University and the University of California who obtained the so-called Cohen-Boyer patents on recombinant DNA technology. |
| Sim, B. Kim Lee | Molecular biologist who is the executive vice president for process development and manufacturing at Sanaria Inc., the company that has developed successful vaccines against malaria. |
| Smith, Edward | Judge, Court of Appeals for the Federal Circuit (1982–89). |
| Swanson, Robert | One of the cofounders of Genentech in 1976. |
| Sweet, Robert | Federal judge in the Southern District of New York who in 2010 decided the first instance of the case *Association for Molecular Pathology v. USPTO.* |

| | |
|---|---|
| Szostak, Jack | Professor (1981), Department of Molecular Biology at MGH; Nobel Prize winner (2009). |
| Theiler, Max | South African virologist who in 1937 developed a yellow fever vaccine and tried it on himself. |
| Tsui, Lap-Chee | Scientist at the University of Toronto who, together with Francis Collins, first identified and isolated the gene for cystic fibrosis. |
| Tyree, Lucas | Member of the Monacan Tribe of Virginia; inventor of foliar feeding formulations. |
| Wallace, Henry | Vice-president of the United States and founder (1926) of what would become Pioneer Hi-Bred International. |
| Wands, Jack | Until his passing in 2023, Dr. Wands was a professor of gastroenterology and medical science at the Alpert Medical School of Brown University. As a researcher at Massachusetts General Hospital in 1988, he successfully appealed an adverse decision on one of his applications from the Patent Office to the court of appeals. His case, known as In re Wands, became—and to this day remains—a fundamental precedent in US biotechnology patent law. |
| Watson, James | Scientist at MRC in Cambridge who in 1953, together with Francis Crick and Rosalyn Franklin, elucidated the double helix structure of DNA. |
| Westheimer, Frank | Professor of chemistry at Harvard and the author's PhD advisor (1912–2007). |
| Winter, Greg | Molecular biologist who at MRC made antibodies by recombinant DNA and in 1989 founded Cambridge Antibody Technologies. |
| Yee, Jiing-Kuan | Scientist at UC San Diego School of Medicine who, together with Theodore Friedmann, challenged UC to be included as co-inventors in a patent on gene therapy. |

| | |
|---|---|
| Zhang, Feng | Professor at Broad Institute of MIT and Harvard who has contested the priority of the invention of CRISPR when carried out in mammalian cells |

# Introduction

More than fifty years ago, in 1971, the late professor Paul Berg, then chair of the Biochemistry Department at Stanford University, combined the DNAs of two different viruses to create the first so-called recombinant DNA. In the process of doing so, he generated an entirely new and artificial virus and, wittingly or not, opened the doors to a transformation of biology that would change the world. The new biology was marked by the ability of humans to modify nature in previously unimaginable ways.

Soon after Berg's announcement, scientists started reprogramming other organisms by using recombinant DNA technology, so that the creatures they altered would do things that they could not do in nature. They made bacteria that produced human insulin, cotton plants that were resistant to pestilent insects, and goats that discharged spider silk into their milk. In short order, bookish biologists were being courted by venture capitalists and multinational companies. And researchers went from being merely curious about how the world functioned to trying to profit from their newfound recognition. As a result, pure biology developed into commercial biology and the participants became interested in monetizing it. And as biologists were inventing all sorts of newfangled organisms, they became acquainted with patents.

The gap between generating essential biological knowledge at universities or other research centers and creating commercial products is hard to surmount without patents. Absent patent protection to give the pioneers a few years of breathing room in the form of exclusivity, most of the work involved in bringing innovative products to the public would not get done. Abraham Lincoln, an inventor in his own right and one of our most eloquent presidents, had high praise for the US patent system.[1] In 1858 he commented that patents "added the fuel of interest to the fire of genius."[2]

I am a patent attorney. For four decades, I have represented clients from all walks of commercial biology, whether individuals, companies, or universities. I have advocated for a plethora of technologies, whether relatively small biological improvements or Nobel Prize–winning inventions.

The legal, commercial, and social debates surrounding the patenting of biological materials—whether living single cells, whole plants or animals, or the inert molecules which they produce or from which they are made—are the subjects of this book.

As biology became marketable, biologists—whether willingly or dragging their feet—became familiar with the basics of patent law. I was no different. In 1976 I received a PhD at Harvard after defending a research thesis on the chemistry of enzymes, a theme with a distinctly biological bent. At the end of my education and training, I became aware that biologists and biochemists were starting to move into the commercial world, talking to investors, and forming their own startups. These developments fascinated me. So, instead of an academic position, I went to law school and, in 1980, was two years away from becoming an attorney. I was in law school at the dawn of the new era. Nineteen eighty was the year that the US Supreme Court decided, in *Diamond v. Chakrabarty*, that, notwithstanding their aliveness, genetically engineered microbes were patentable.[3] That caught my everlasting attention.

Upon graduation, I became one of a handful of patent attorneys in the United States with training in biochemistry and molecular biology, the subjects of my research work. I mounted a surfboard as the wave of biotechnology inventions was rising, and I have been riding it since, so far without crashing. I have witnessed and participated in some of the landmark intellectual property (IP) decisions of the last forty years in this remarkable area of technology and law.

Through the unique lens of IP, I will take you on a voyage that spans more than half a century, from the pioneering experiment of Paul Berg in 1971 to build the first recombinant DNA virus, to the now-routine genetic engineering of plants and animals, to the latest DNA-editing techniques such as CRISPR, which promise to cure inherited genetic diseases in humans. Our travels have five stopovers:

- We start with the invention of patents in fifteenth-century Venice. From there we move on to the twentieth century, especially the 1970s. That was the decade that marked the ambivalent beginnings of commercial biology.
- In the years that followed, the courts learned to deal with the many questions that arose as classical IP law was forced to adapt to an onslaught of groundbreaking developments. They struggled with issues such as, Who owns what is in your body and in your mind? Can you patent living beings? How do you legally describe life?
- I will apply the lessons learned from the courts to a few of my specific cases. They deal with patenting live beings, from small microbes to large cows. In doing so, we go to the courtrooms of the Court of Appeals for the Federal Circuit, so you can witness some of my hearings.
- If left unattended, the patent system can become unbalanced, and it can then use some healthy criticism. In this stopover, I give voice to those who question if the system is working well or if it has become distorted over the decades. We explore things like the high prices and availability of patented drugs, especially biologics; concerns about lack of access to the latest vaccines; and the conquest of world agriculture by genetically modified crops.
- At the end of our voyage, and inspired by the Hebrew concept of *tikkun olam,* "to repair the world," we consider the possibility of using the patent system to bring economic and social improvements to the lives of disenfranchised Indigenous communities.

Because of my focus on IP, the stories I tell you are grounded in science and law. It is impossible to sidestep these two disciplines in writing a book like this. While both fields can be impenetrable with jargon and opacity, I have tried very hard to make them more transparent and accessible to nonspecialists.

The main text is written in broad strokes and tries not to "get into the weeds" too much. However, throughout the chapters I have added endnotes that explain things in more detail. When discussing IP policy,

I have included endnotes with opposite opinions, so as to provide depth and balance to the argument. Readers interested in the weeds may then dig deeper.

But beyond science or law, my tales have human faces. They are the faces of the entrepreneurs, scientists, lawyers, and judges involved in the rise of commercial biology and its intellectual property. The protagonists of these tales and the legal battles that I fought on their behalf are at the core of this book.

Bethesda, Maryland, 2024

# Part I

# The Birth of Commercial Biology

Let's start with the scientific and technological revolution that occurred in the wake of the late professor Paul Berg's combining the DNAs of two different viruses to create the first so-called recombinant DNA. He did that in his laboratories at Stanford University. Since then, universities throughout the world have played an outsized role at the beginning and in the development of the new science. We will explore how the commercialization of biology changed the world of academia. And we will see that by bringing many more women attorneys into intellectual property careers than before, the new biology also changed the legal profession.

# 1

# The Invention of Patents

Because of the critical role that intellectual property plays in linking pure science to commercial applications, I will first give you a short introduction to the concept of patents. A good way to understand the idea of a patent is to go back more than five hundred years to the fall of Constantinople. It was in the wake of that city's fall that the neighboring Venetians invented the basic notion of our modern-day patent systems: exclusivity in exchange for disclosure.

## Traveling Artisans

Constantinople, the commercial jewel of the Old Byzantine Empire, fell to the armies of Sultan Mehmed II in May of 1453. The inhabitants who didn't escape before or during the pillaging were allowed to return to their homes. Yet plenty of them decided not to stay. The next decades saw the emigration of many survivors, among them highly skilled artisans, craftsmen, and engineers. The fall of Constantinople capped two centuries of migration of Byzantine artists and merchants to Venice, in the wake of ever-encroaching Ottoman conquest. Lots of silk weavers, canal makers, and glass blowers ended up in Venice.

The arrival of such skilled immigrants did not go unnoticed to the doge and his council. Following earlier isolated instances of regulating the professions, in 1474 the Venetians enacted a clever statute by which the traveling artisans would be encouraged to stay instead of moving on to northern Europe. This was the Venetian Patent Statute, considered to be the first comprehensive patent law of the Western World (see figure 1.1).[1]

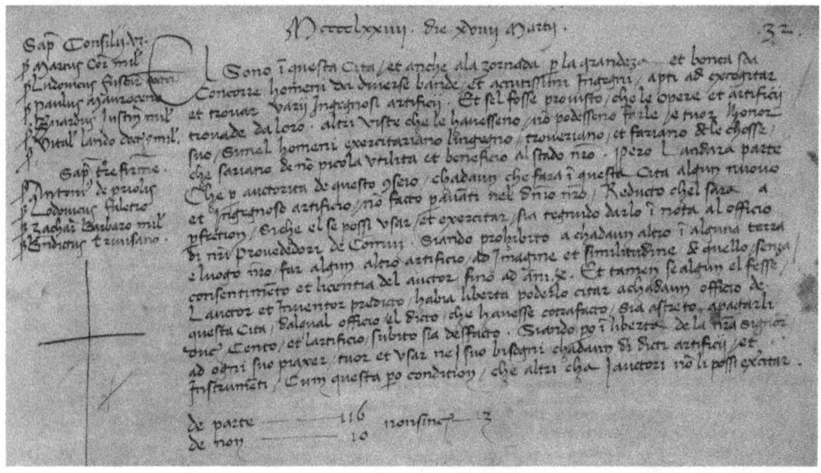

Figure 1.1. A reproduction of the Venetian Patent Statute of 1474. (Figure from the Senate of Venice.) Reproduced from the public domain.

Like all legislation before and since, the law starts with a description of its intent, in this case, embellished by lighthearted pride in La Serenissima, the Most Serene Republic of Venice: "There are in this city, and also there come temporarily by reason of its greatness and goodness, men from different places and most clever minds, capable of devising and inventing all manner of ingenious contrivances. And . . . men of such kind would exert their minds, invent and make things which would be of no small utility and benefit to our State."

Then comes the denser legalese, with interminable sentences and enough "heretofore"s, "aforesaid"s, and "thereof"s to make a lawyer blush:

> [By] authority of this Council, each person who will make in this city any new ingenious contrivance, not made heretofore in our dominion, as soon as it is reduced to perfection, so that it can be used and exercised, shall give notice of the same to the office of our Provisioners of Common. It being forbidden to any other in any territory and place of ours to make any other contrivance in the form and resemblance thereof, without the consent and license of the author

> up to ten years. And, however, should anybody make it, the aforesaid author and inventor will have the liberty to cite him before any office of this city, by which office the aforesaid who shall infringe be forced to pay him the sum of one hundred ducats and the contrivance immediately destroyed.[2]

The basic idea was simple: Stay in Venice, teach us what you know, and we'll enforce it, so no one else can copy your stuff for ten years. But beware, ye of most clever minds! If you want exclusivity, you must invent a "new and ingenious contrivance." You are not getting exclusivity for contrivances that we already know about. No, Dear Sir, your invention must not only be novel, but it must be ingenious; that is, "non-obvious" (as modern United States law has it), or "inventive" (using the words of contemporary European law). Your invention "must be reduced to perfection," meaning that it must be built, not just prophetic. And it must be "of no small utility" to Venice; that is, it must be "useful" (as in US law) or have "industrial applicability" (as in European law). You must describe your invention "so that it can be used and exercised." In other words, it must be workable and reproducible by others. If it can't be "exercised," we will know that you have kept secrets from us, and *that*, if we may say so, is not a good idea. If you are even thinking of cheating us, Dear Sir, just take a good look at the Bridge of Sighs, which leads to prison.

You "shall give notice to . . . the office of our Provisioner," means, in modern parlance, that the inventor must file a patent application with the national patent office. If an infringer appears, then his copied contrivance "will be immediately destroyed." Even that drastic result happens nowadays: If the court directs an infringer to stop, it may sometimes order his infringing products destroyed. And, just as today, the Venetian Senate could give money damages to the wronged patent owner.

The core of every present-day patent system has the basic quid pro quo of the Venetian Statute: limited exclusivity in time and place in exchange for full disclosure of the invention. And pretty much every country in the world has a patent system with Venetian features: from communist China, through socialist Sweden, to our unashamedly capitalistic United

States. Small African countries have it, and the large Russian Federation has it.

The Venetian idea traveled well, reaching France, England, and eventually crossing the Atlantic. The United States Constitution of 1789 gave the Congress the "Power to promote the Progress of . . . the useful Arts, by securing for limited Times to . . . Inventors the exclusive Right to their respective . . . Discoveries."[3] This is known as the Patent Clause. In 1790 the Congress enacted the first US Patent Statute, giving the ingenious inventors of the new republic the patents that they deserved.[4]

Thomas Jefferson, a prolific inventor in his own right, was appointed as one of the first members of the United States Patent Board. He was, in essence, a patent examiner and he eventually grew tired of the job, moving on to become secretary of state, vice president, and, in 1801, the third president of the United States. Another famous patent examiner, Albert Einstein, sitting in the Swiss Patent Office in the early 1900s, also got bored with his job and, while daydreaming at his desk, formulated the theory of relativity. Examining patent applications seems to be a harbinger of larger ambitions.

In a much-quoted letter of 1813, Jefferson described the patent right as an "embarrassment" but one worth having: "I know well the difficulty of drawing a line between the things which are worth to the public the embarrassment of an exclusive patent, and those which are not."[5] He seemed to imply that he was not crazy about granting exclusivity but he might go along if it was for the public good.

Jefferson's careful balancing act has resonated through the centuries.[6] The debates about whether patents are good or bad found echo in Professor Berg at Stanford. Berg, you may remember, created the first "recombinant DNA" organism in 1971, by combining the DNAs of two different viruses. But he did not file for patents. He saw himself as a pure biologist, interested only in science and not in commerce.[7]

And the debate continues to this day in never-ending battles. How much patent protection should we give to the cutting-edge "useful Arts," such as genetically engineered bacteria, plants, or animals? And what should we do when a private company holds a patent on a genetic vaccine much needed to stop a global pandemic? Should we even let it hold such a patent? Why are patented drugs so pricey in the United States, and can we do anything about it? These are all timely questions to which we will soon turn.

## A Generational Gift

The patent systems of the world try to influence decisions made by competitors in the private sector. Biotech companies that sell drugs to the public much prefer to keep secret the chemical processes, purification methods, or genetically engineered bacteria used to produce their medications. In contrast, we the public would like the companies to disclose these valuable secrets, not just the final medications. That way, the secrets can be available for further research and use. Disclosure of technology is the nourishment of further invention.

Just like the Venetians did in the fifteenth century, governments that enact patent laws give a company limited exclusivity as an incentive to disclose secrets. About a year and a half after filing, the US Patent Office publishes pretty much all pending patent applications.[8] The public then has access to the (no-longer) secrets and can use them to do basic research with them, and on them. Nobody but the patent owner, however, can *commercialize* the invention when the patent is granted a few years after its publication. And that prohibition continues until the patent expires, a little less than twenty years later.

The processes and sources described by a patent application must meet basic legal requirements: they must be novel, non-obvious, useful, well described, and reproducible. If such is the case, the company can get its patent exclusivity but only for a while and only on a nation-by-nation basis. With a few exceptions there are no international patents, only national ones. After the patent term ends in any one country, its sources and processes are commercially usable by everyone. At such time, the invention no longer belongs to anyone and can be freely used by all competitors.

In 1790 the United States Patent Board issued three patents. In 2021 the yearly number was up to 374,000. In May of 2021 the Patent Office issued patent 11,000,000; and we are now well over that milestone.[9] About 7,000 US patents issue every week and, after subtracting premature expirations because of lack of interest or obsolescence, they expire about eighteen years later. That is thousands of patented technologies coming into the public domain every year.

Randall Rader, who between 1990 and 2014 was a sitting judge of the Court of Appeals for the Federal Circuit, the appellate court that hears and decides patent cases in the United States, described the limited time of patent exclusivity with an insightful comment: "It's a generational gift," I once heard him say at a lecture.[10] "A generation invents useful things; we let them have the things for themselves for about two decades, and after that, their inventions go to the next generation; for free." And he finished with a smile, "Not a bad deal."

Indeed, it is a good deal for both sides. The inventive companies remain free of competition for a while, hopefully long enough to recover their investments and start making a profit. We, the public, get the almost immediate disclosure of new technologies that would otherwise remain secret. Then, eighteen years later, give or take, the companies' competitors get to copy the technologies and may start profiting from them. Competition brings prices down, and we, the public, benefit.

Another, subtler, consequence of forbidding the use of patented technology for a while is that it forces competitors to get around the patent. This is known in the jargon as "designing around" a patent. A good, albeit inverse, analogy might be the antibiotic resistance of bacteria. A bacterium whose survival is blocked by an antibiotic may evolve to one that is antibiotic resistant. It then lives on, in a vicious cycle that bewilders physicians. In similar Darwinian fashion, technology that is blocked by a patent will oftentimes lead to new inventions, some of them patentable in their own right. This generates a virtuous cycle of leapfrogging innovations that delights all, especially the patent lawyers.

A classic example of how the generational gift works is the different pricing between patented medicines and their generic versions. The former are usually expensive, sometimes too much so, while the prices of the latter are dramatically lower. Take for instance the antidepressant

Zoloft®.[11] It was invented and patented by Pfizer in the 1970s and approved by the Food and Drug Administration of the United States (FDA) in 1991. The Pfizer patent expired in 2006. By the expiration date, Ivax Pharmaceuticals had already obtained approval from the FDA to immediately start selling the generic version, sertraline. The price of generic sertraline today is about 5 percent that of Zoloft. The generic allows consumers (and their insurance companies) to save thousands of dollars a year in prescription costs.[12]

Of course, it bears remembering that without Pfizer's investment in antidepressant research, chances are there would have been no Zoloft; without a patent system there would likely not have been a Pfizer investment and no commercially approved drug. And, without Zoloft, there would likely not have been a generic sertraline. Companies like Pfizer must invest in research and development in the first place. They do so only when their legal departments tell them that they will have patent exclusivity for a while. And at the end of the road, usually around two decades later, there is always an Ivax waiting for the generational gift. The whole system is designed to be a careful balancing act.

# 2

# The Path from Pure to Commercial

The patent system played a fundamental role in the transition from pure to commercial biology that the world experienced in the last fifty years. But taming nature to make it work to humanity's advantage did not start in the 1970s. It is much older.

For most of recorded history, inventive humans have been tinkering with Mother Nature. Twelve thousand years ago, the people of the Fertile Crescent domesticated barley and started the agricultural revolution.[1] Animal breeders for centuries played roulette with the genetics of their dogs, hoping to get bigger, cuter, or faster canines. A century ago, plant breeders created hybrid corn plants with a vigor that its two individual parents did not have. Until recently, however, no one really understood what was going on inside the barley or the dogs.

## From Randomness to Rigor to Business

Gregor Mendel, a mid-nineteenth-century Augustinian monk, tried to understand. He was one of the first scientists to move biology from the mere observational to the rigorous. While doing fundamental research on pea breeding in his monastery in Brno, Czech Republic, Mendel discovered recessive and dominant "factors," which we now call "genes."[2] Because of his groundbreaking investigations, Mendel is known as "the father of genetics."

If Mendel were able to time travel to our present and meet a twenty-first-century geneticist, he would be delighted to find that we now have the science to explain to him in detail what his peas were doing as he crossbred them in the gardens of his monastery. Instead of random mixing, breeding, growing, and then hoping for the best, we now can do precise genetic surgery on the microbes, plants, and animals of our choice. We have come a long way in removing the age-old element of randomness from the whole process. The new biology, whether it edits genes or modifies proteins, is no longer the black box of old, but the brightly illuminated science of our new age.

Yet mere understanding of how the biological world functions was never enough. We humans have always been intent on taking advantage of nature's resources to improve our lives, especially the things we eat and drink. Biotechnology, which may be defined as the use of biological materials in industry, is not new. Long ago, we learned, probably by serendipity, that yeast spores blowing in the wind could be recruited for fermenting beer and leavening bread. Archaeologists have found 13,000-year-old evidence of beer brewing in the Carmel Mountains near Haifa, Israel. Pliny the Elder, the Roman historian, wrote of Iberians skimming foam from beer production to leaven their bread. The Chinese made wine 9,000 years ago.[3] These initial uses of yeast for making foods show that biotechnology is as ancient as the hills.

Beyond enlisting nature to better our lives, however, there is a major difference between today's biotechnology and that of the past. In the last half century, there has been an explosive commercialization of biology in which discovery is no longer solely pursued to attain knowledge or make random improvements to seed or stock. If Mendel arrived on the biotech scene today, he would be taken aback to find that tens of thousands of companies have arisen to profit from our increased understanding of his genetic "factors."

The path from basic research to biology as business started in academic labs during a few years in the mid-1970s. It was a time of great intellectual excitement but also deep ambivalence, as university scientists and their administrators came upon the crossroads that separated pure from applied science. Some hesitantly crossed over the boundary while others got lost at the intersection.

## Gene Splicing

The fundamental events that led to commercial biology occurred in one single decade, the '70s, and mostly in California. The events had to do with the creation of hybrid organisms.

As a prelude, in 1953 came the elucidation, by James Watson and Francis Crick (with the then unrecognized contributions of Rosalind Franklin), of the double helix structure of DNA.[4] It happened in the labs of the Medical Research Council (MRC), at the University of Cambridge, in England, one of the great research centers of the world. Their moment of insight crystallized the dual concept of DNA: a molecule with a well-understood chemical structure as well as a repository of genetic information. Like an architect's blueprint, DNA carries the chemical instructions upon which all living beings are built. The duality of DNA as both a tangible molecule and as intangible information would eventually lead to profound legal and philosophical debates.

In the 1960s scientists came to understand that the genetic code was universal, meaning that the blueprint for a human being was written in its DNA in the same alphabet as that for a lowly bacterium. This meant that one could possibly transfer a human gene into a bacterium, and the tiny recipient would immediately recognize and read the DNA from the donor. This would be like an architect in Mumbai building a house identical to her colleagues' house in Miami when both are working from the same blueprint. Just like these transnational architects share the same architectural grammar, bacterial and human machineries speak the same DNA language. Without a universal genetic language, nothing would have happened the way it did. Universality allowed all future developments, especially the ability to move genes from one organism to another.

As we have seen, in 1971, at the threshold of the decade that led to commercial biology, Paul Berg at Stanford created the first recombinant DNA. This hybrid DNA carried the blueprints from two distinct viruses and resulted in a chimeric virus. The construction of Berg's chimeric virus was what might best be called a "proof of principle." It proudly said to the world, "Look what I have done!" Indeed, in his paper in the *Proceedings of the National Academy of Sciences* describing his momentous experiment, Berg admitted that he had no idea what this new creature could do. He just said, "We have no information concerning the activities of the [chimeric virus], but appropriate experiments are in progress."[5]

Worried about what he had just begat, however, Berg stopped further progress. He became concerned that creating additional interspecies chimeras would lead to runaway problems, such as the spread of cancers. In 1975, in a remarkable demonstration of scientific self-restraint, Berg and his colleagues organized the Asilomar Conference on Recombinant DNA.[6] The conference included an interdisciplinary gathering of biologists, lawyers, and doctors. It issued a set of guidelines on further genetic engineering research, putting serious restrictions on working with biological hazards. Conference attendees feared that reckless researchers would genetically engineer deadly toxins or harmful agents and contaminate the environment. Henceforth most academic work on recombinant DNA would willingly comply with the Asilomar guidelines. Although these guidelines regulated basic research in the new science, they couldn't prohibit commercial development. This limitation would prove to be a critical argument in favor of patenting the basic techniques of genetic engineering by Stanford and the University of California.

Paul Berg, a philosophically "pure" scientist, was at the far side of a timeline that divided the biological research world in two. On the earlier side were those who did research for the sake of acquiring knowledge. On the later side were those who did it so that it could be turned into money; that is, so that it could be "monetized." Berg did not think commercially. That had to wait a few more years.

In 1973 Stanley Cohen at Stanford and Herbert Boyer at the University of California started collaborating on so-called restriction enzymes.[7] These are tiny molecular scissors that cut DNA from whatever source—whether bacterial, vegetable, or animal—with surgical accuracy. Doing so allowed them to cut open the DNA from an organism of one species and, at a specific spot, splice in a piece of DNA from an organism of a different species. It was like cutting a long string of white pearls from Japan at a precise place, inserting there a fragment of blue pearls from

Tahiti, closing the loop, and creating a gorgeous white necklace with a blue segment. This work led to the astonishing creation of an interspecies bacterium that carried a frog gene and produced a frog protein. The frog protein was just an appetizer; it was not yet pharmaceutically useful. Because of these experiments and the patents that they obtained, Cohen and Boyer are considered to be the inventors of genetic engineering.

In 1976 Herb Boyer, the UC professor, together with Robert Swanson, an alumnus of the San Francisco venture capital firm of Kleiner, Perkins—a firm renowned for taking high risks with its money—founded Genentech. The birth of commercial biology as we know it today was the brainstorm of two men from very different worlds. One came from molecular biology and the other from venture capital. Chemistry and money had been intimates since the nineteenth century industrial production of dyestuffs and of synthetic drugs and plastics. Physics and money had also merged very early, with the steam engine, electricity, and optics. Genentech was the first of its kind: it was an interspecies baby from the marriage of biology and money.

The first splicing of a gene for a useful product occurred in 1977, a year after the founding of Genentech. The gene was that for human somatostatin, a natural molecule involved in regulating the hormonal system.[8] In 1978 another genetically engineered bacterium produced human insulin.[9] These proteins were made by Arthur Riggs and Keiichi Itakura, two scientists at City of Hope National Medical Center in Los Angeles, using the new recombinant DNA technology of Cohen and Boyer. Their work was the result of an arrangement between City of Hope and the brand-new Genentech. Genentech paid City of Hope for making the proteins and, in return, controlled the patents and owned the molecules; that is, it controlled the intellectual and tangible properties.

Biologists had been walking down the path of research for centuries, and in 1976 they came upon a road that they would have to cross. On one side of the road was the world of pure research. It was the world of Gregor Mendel, James Watson, and Paul Berg. If they crossed it to the other side, they would get to the land of monetized biology. That was the land of Stan Cohen, Herbert Boyer, and Genentech. Some crossed it, some stayed where they were, and some got lost at the juncture. The founding of Genentech marked the earliest of the crossings.

It is hard to fathom which "Eureka!" moment was more profound: whether the recognition by Cohen and Boyer in 1973 that their little engineered bacterium was churning out a frog protein, or whether it was the realization in 1976 by Boyer and Swanson that they could commercialize genetic engineering, give new drugs to the world, and become multimillionaires in the process. Regardless, a major paradigm shift occurred in the short few years spanning the middle '70s. The world was forever changed.

## Magic Bullets

There was another biology breakthrough during that prolific decade: the creation of monoclonal antibodies. This effort was led by César Milstein, an Argentine-born immunologist, and George Köhler, a German biologist, working together at the University of Cambridge.[10] Their laboratory was also in the MRC, the same place where Watson and Crick, twenty years earlier, had figured out the structure of DNA. Milstein and Köhler's fundamental work would eventually lead to the creation of so-called pharmaceutical magic bullets, highly selective drugs for the treatment of disease.

The dream of developing magic bullets in the form of antibodies that would cure sickness with minimal side effects goes back to Berlin immunologist Paul Ehrlich, who coined the phrase in the early twentieth century.[11] In 1975 Milstein and Köhler invented a way to create highly selective antibodies able to play their part in Ehrlich's dream. The scientists fused two kinds of cells: immortal cancer cells and perishable immune B-cells, which produce antibodies in our bodies. The resulting chimeras, known as hybridomas, have traits from each of their parents: they are immune cells that have become immortal. And, because it is possible to "clone" or isolate from a mix of hybridoma cells a single cell that produces a single type of antibody against a disease agent and not the complex mixture normally obtained from regular immune B-cells, the antibodies are known as *monoclonal* antibodies.[12] These would eventually become the magic bullets of Ehrlich's vision.

Hybridoma cells, like the genetically engineered bacteria that Cohen and Boyer were creating near San Francisco, are also hybrid forms of life. And, just as the joint inspiration of Boyer and Swanson in 1976 gave birth to Genentech, the creation of hybridomas led, in 1978, to Hybritech, a company founded for the purpose of exploiting the commercial possibilities of the new immunology. Like Genentech, Hybritech was also the brainstorm of biologists and moneymen. But unlike Boyer, César Milstein and his colleagues had nothing to do with it. That fell upon scientists of the University of California in San Diego. To found Hybritech they received seed money from Kleiner, Perkins, the same risk-taking venture capital firm that funded the early years of Genentech.[13]

Hybritech was not in England, the site of the MRC where Milstein and Köhler worked, but in La Jolla. The early riches of the new biotechnology were all in an extended West Coast family: California scientists creating California companies funded by California capitalists.[14] The family did not include Milstein and Köhler, the Cambridge brains behind monoclonal antibodies. These two pioneers were left out of the California dreams and their treasures. Cohen and Boyer, aided by prescient university officers and venture capitalists, walked over the commercial intersection to the world of monetized biology. The MRC scientists, however, got lost at the crossing. There was no one knowledgeable enough waiting there to lead them to the other side. The guides at the intersection were both ambivalent and ignorant, a deadly combination. This is a sorry tale of patents in academia that I will discuss in chapter 3, after I tell you about the explosive growth of commercial biology.

## A Meadow Full of Dandelions

After the creation of Genentech and Hybritech, biotech companies sprouted at the other side of the road like dandelions on a hilly meadow. When the wind picked up, many of them, in turn, dispersed their seeds to the rest of the hills, as spinoffs and subsidiaries. And when the big pharma companies realized how much gold there was in "them thar hills," they started gobbling up the dandelions, becoming giant biotech companies themselves. If 1976 marks the first edge in the culture shift in biology, 1980 marks the other side, the beginning of a new era. Starting that year, the movement of genes from one organism to another quickly became, and is today, big business.

In the 1980s Genentech took off like a rocket, although its birth was not free from difficulties, as is sometimes the case with births. The company partnered with Eli Lilly to commercialize Humulin®, the brand name for human insulin made by genetic engineering. After it received approval from the FDA in 1982, Humulin became the first "recombinant" drug sold in the world. Lilly was still selling it in 2023. This highly successful drug was the direct descendant of the 1976 contract between Genentech and City of Hope. City of Hope complained for years that Genentech was not paying it sufficient royalties on the sales of Humulin. It eventually tired of asking and, in 1999, sued Genentech for breach of contract. After several trials and appeals, in 2008 Genentech was ordered to pay City of Hope close to $300 million in overdue royalties.[15]

This sizeable monetary award did not slow Genentech, and to this day the company continues to invent ever more sophisticated recombinant drugs for the treatment of growth hormone deficiencies, acute heart attacks, lymphomas, breast cancer, and viral infections. In 2009 Genentech became a wholly owned subsidiary of Hoffman-LaRoche AG, a Swiss multinational pharma company, which paid $47 billion for it. That's billion with a B.

Genentech was not the first biotech startup that was eventually wolfed down by a multinational pharma company. There are many others, and in different areas of the new biology, whether they are in the business of modifying plants or animals or of making magic bullets.

In the 1980s, Plant Genetic Systems (PGS), of Ghent, Belgium, using the new genetic engineering, was the first to harness a naturally occurring biological phenomenon to generate pest-resistant plants. Its scientists extracted from a bacillus living in the gut of insects the gene for a pesticidal protein called *Bt*. In nature, the bacillus uses the *Bt* protein as ammunition to weaken its host insects. The insects, in turn, are natural plant foes, feeding on cotton or corn. PGS scientists transferred the *Bt* gene of the bacillus directly into the target plant, allowing it to defend against its predatory insects by using the bacterial ammo.[16] Think about that: genetic engineers recruited the *Bt* bacterium, a natural enemy of the insect, and co-opted its toxins to shield a plant from the very insect that is a common adversary to both the plant and the bacillus.

The concept of protecting plants using ammunition from the enemy of their enemy proved enticing enough to others that, like bollworms feeding on cotton, PGS was promptly gobbled up. In an almost biblical series of begettings, PGS was acquired by AgrEvo, which then merged with Rhone Poulenc-Agro to form Aventis Crop Science, which was bought by Bayer to form Bayer Crop Science, which is now one of the three major divisions of Bayer AG of Leverkusen, Germany.[17] The names of the successive gobblers are not as important as the stocking of the biotech pond with ever bigger and hungrier fish. The *Bt* gene, first extracted from its tiny bacillus in Ghent, went along for the ride.

The monoclonal antibodies invented by Milstein and Köhler at the MRC in England led to the Hybritechs of the world, which sold antibodies mainly for doing clinical blood tests. These tests are carried out outside the body to diagnose the blood levels of a myriad of conditions and diseases. The real dream of creating magic bullets that would function as therapeutic drugs, however, required injecting them into a human. This was problematic. Since Milstein's monoclonals of 1975 were derived from mice, they were, when given to a human patient, rejected by her immune system as foreign before they had a chance to do their magic.

The dream of Ehrlich's therapeutically magic bullets did not happen until 1986, a decade later. That year, Gregory Winter, a molecular biologist also working at MRC, combined the science of genetic engineering of the Californians, Cohen and Boyer, with the hybridoma science of the Cantabrigians, Milstein and Köhler. Winter transferred antibody genes into bacteria and produced the first genetically engineered antibodies. Then he went further. He invented a way to make his antibodies humanlike, and "humanized" the mouse-derived monoclonal antibodies of Milstein.[18] Since these antibodies were now humanlike, they would no longer be rejected by the human system. There must be something in the water of the MRC that leads to such great insights.

In 1989 Winter founded Cambridge Antibody Technologies (CAT). There, together with Abbott Labs, they made Humira®, a treatment for rheumatoid arthritis. Humira was the first fully human monoclonal antibody useful for injection. Humira was launched in 2003. After Abbott split in two in 2013, AbbVie, one of its two progeny, inherited Humira. By 2017 Humira had market sales of $18 billion. If Paul Ehrlich could time-travel to the present, he would shed tears of joy at contemplating the marvelous Humira, the first useful magic bullet of his dreams. Oh, and Greg Winter and his investors sold CAT to AstraZeneca in 2006 for close to a billion dollars. Talk about monetizing biology.

There are now tens of thousands of biotech companies around the world, many of them whispering privately that they will be a reincarnation of Genentech or CAT. They are led by scientist-entrepreneurs who dream of being the next Herb Boyer or Greg Winter. The US market for biotechnology in 2023 was estimated to be close to $193 billion.[19] It is likely that the worldwide market was close to ten times as much, a cool 2 trillion.

All of this is thanks to two things. One is the universality of the genetic code, which allows genes to be surgically inserted into places where they have never been before. The other is the rude insertion of patents into the placid world of academic biology.

# 3

# A Clash of Two Worlds

It was in 1974 that the world of pure biology merged with the world of patents. In May of that year, Niels Reimers, a member of Stanford University's Sponsored Projects Office, was sitting in his office reading the *New York Times*. He ran upon a story reporting that some important work had come out of Professor Stanley Cohen's lab, a few blocks away. You may recall that a year earlier, Stan Cohen at Stanford and Herb Boyer at the University of California (UC) had started collaborating on restriction enzymes. With these enzymes, the two scientists had created a bacterium that carried and produced a frog protein. In November 1973 Cohen and Boyer published their work in the *Proceedings of the National Academy of Sciences, USA*. This was the first paper ever on modifying bacteria with recombinant DNA that contained genes from other organisms.[1] The last paragraph of the paper predicted cautiously, as is the wont of most careful scientists, that the procedure could be potentially useful for transferring DNAs from other species. An astute science journalist for the *New York Times* had read this paper and reported on it.[2]

Having worked in industry before arriving at Stanford, and trained to think about patents, Reimers picked up the phone and called Cohen. When I interviewed Reimers in 2023, he remembered the moment as though fifty years had not passed. "'Stan, this looks like very important work,'" he recalled saying, and he added to me, "Stan agreed but told me he did not want to patent it. He wanted it widely disseminated."[3]

After a few gives and takes, Reimers convinced Cohen that patenting the techniques of recombinant DNA might not be a bad idea after all. He argued that the university might make money and use it to fund research at Cohen's lab. He added that obtaining patents would allow Stanford to give out licenses to the patent on the condition that industry comply with the Asilomar Guidelines.[4] The guidelines placed restrictions on work with genetically engineering hazardous materials. Private firms were not bound by the guidelines, as these only applied to work funded by federal agencies, such as the National Institutes of Health (NIH). Asilomar did not restrict firms that did not receive federal money. Yet, explained Reimers to Cohen, getting Stanford and UC a few patents on recombinant DNA technology and licensing them to industry with Asilomar restrictions written into the contracts would bring industry into the fold. It was a brilliant argument, and the originally reticent Cohen agreed.

By the time of their conversation, it was too late to get any patents other than in the United States. The publication in the *Proceedings* had destroyed patent rights everywhere else. The United States had and still has a one-year grace period from the date of publication of an invention for filing an application that results in a US patent. That gave Reimers about six months to get going before losing US patent rights. He called Bert Rowland, a West Coast patent lawyer with a PhD in organic chemistry.

Rowland, a smart man with a terrific imagination, immediately understood the long-term commercial implications of recombinant DNA technology. He wrote and filed on time a patent application that years later resulted in three fundamental patents on the new genetic engineering. They are known as the "Cohen-Boyer patents." These patents, which issued around 1980, dominated the commercialization of the new science for the next twenty years.[5] At that time, US patents expired seventeen years from the date they issued. During the seventeen years after issuance, the Cohen-Boyer patents made Stanford and UC a lot of money.

## Initial Ambivalence

It should be obvious that modern commercial biology is not as much the product of private companies as it is the result of the collaboration between university and company labs. The scientists responsible for many of the early breakthroughs, which led to the surge of biotech startups, came from academia: Stanford, UC San Francisco, MRC in Cambridge, City of Hope, the University of Ghent. Today, the relation between academia and the biotech industry is intimate. It was not always that way.

Let's go back to splitting the space-time of the biological research universe into two: the academic territory before the founding of Genentech in 1976, and the commercial territory after 1980, when the monetization of biology lost all inhibitions. In contrast to the chemical and engineering worlds, the realm of university biologists was, until the mid-1970s, centered on basic research. Most biologists tinkered with how cells worked, what DNA did, how it led to proteins, and how these proteins folded into functioning molecules such as enzymes. After they completed their research, they published it and, when asked, freely gave away their biological materials to colleagues around the world. While their contemporaries in the chemistry labs across the university quad had long consorted with industry, serving as consultants, or even trying to get patents on their inventions, the biologists didn't. Few companies sought biological consultants, and, besides, there was nothing biological to patent.

Paul Berg belonged in the precommercial world. In an interview held with Berg in 1980 on his receiving the Nobel Prize in Chemistry, Berg said about his days of research in the late '60s: "[When] I was growing up in the university, nobody ever asked me to patent anything and I never patented anything in my life. . . . I gave away everything we ever did because what we did it for was to benefit research."[6]

Even the invention of transferring foreign genes into bacteria by Stan Cohen and Herb Boyer might never have been patented were it not for the prescience of Niels Reimers. But not everyone was pleased with the two scientists' IP decision. In 2010, in looking back at the filing and issuance of the Cohen-Boyer patents, Paul Berg had some harsh words for the patents' scope and reach. In a paper published that year, Berg said: "Their [Cohen and Boyer's] claims to commercial ownership of the techniques for cloning all possible DNAs, in all possible vectors, joined in all possible ways, in all possible organisms were dubious, presumptuous, and hubristic."[7] It is highly unusual for one famous scientist to throw such strong words at another, and in public. I can only surmise that Berg's rebuke shows how shaken he must have been seeing his beloved world of biology turn the corner into the roads of commerce.[8]

Yet by the time the Cohen-Boyer patents expired in 1997, the entire US genetic engineering industry had taken licenses. The sum of all royalties to Stanford and the University of California, its joint partners, added up to $250–255 million.[9] And that was in the United States alone. Imagine the numbers, had Reimers been told of the gene transfer invention before its publication in May 1973 and still been able to file for patents worldwide. Even so, this kind of money coming into a university from patent licensing was unheard of at the time. It awakened many a university administrator.

Unfortunately for the MRC across the Atlantic, the administrators there did not get the memo. Like Paul Berg and Stan Cohen (before Niels Reimers got to him), César Milstein in Cambridge also belonged to the pure science world. Milstein was vaguely aware of the possibility of patenting biological inventions. With less than a month before the journal *Nature* was to publish his and Köhler's groundbreaking paper on making monoclonal antibodies, he submitted, on July 10, 1975, his still confidential manuscript to the National Research Development Corporation (NRDC). NRDC was the UK entity responsible for patenting inventions coming out of the Medical Research Council. But NRDC took more than a year to respond. Milstein's paper was published on August 7, 1975, destroying all worldwide patent rights, except possibly those in the United States.[10] Had the NRDC woken up within a year of the *Nature* publication, it could have still filed in the United States, just as Niels Reimers did with the Cohen-Boyer patents. Milstein's own publication, if less than a year before a potential patent filing in the United States, would not have prevented the issuance of a US patent.[11] But, because everyone was asleep at the wheel, none of that happened. NRDC was not drinking from the insight-spawning water coolers of the MRC.

There was a lot of finger-pointing in the UK after the failure to file for patents on such a pioneering invention as monoclonal antibodies. The so-called Spinks Report in 1980 blamed the "young scientists" for not seeking patents; that is, it blamed Milstein and Köhler.[12] After reading it,

Margaret Thatcher, who was then prime minister and a chemist by training, was reportedly furious that no patents had been obtained.

Milstein took issue with all those conclusions. In 1989, after being invited to a luncheon with Thatcher, he sent her a dossier of documents attaching copies of correspondence from the critical time. He suggested in his letter to Thatcher that he wanted to lay to rest previous false conclusions regarding his responsibility for the fiasco. Milstein included the response that he had received from the NRDC in October 1976, more than a year after his publication in *Nature*. The NRDC's myopic view of the matter was that there was no "immediate application" for monoclonal antibodies. It could not find any practical applications of a commercial nature. And, covering its institutional buttocks, it implied that even if Milstein's publication had not occurred, patenting his invention would have been nigh impossible. The reason, apparently, was the writers' ignorant opinion that the field of genetic engineering was a difficult area in which to seek patents. There was nothing really patentable in Milstein's 1975 paper, the NRDC concluded.[13] This is what I can only call a doozy of a legal opinion.

The NRDC could have asked an ingenious biotech patent attorney like Bert Rowland, who would have told the corporation differently. My speculation, however, is clouded by hindsight and somewhat unfair to the NRDC. The reality is that in 1975 Rowland was one of few if any biologically trained patent attorneys in the world, and they were—where else?—in California. Most patent lawyers working during the pioneering years of commercial biology were chemists or chemical engineers. In the mid- to late 1970s they were frantically trying to learn biology to understand the soon-to-explode world of patent applications dealing with biotech inventions.

In 1980 Nicholas Wade, a journalist, published an article in *Science* about the monoclonal antibody fiasco. He quoted Milstein as stating, "We were too green and inexperienced on the matter of patents. . . . We were influenced by that psychology [of making methods freely available]. We were mainly concerned with the scientific aspects and not giving particular thought to the commercial applications."[14] During a 1993 meeting on the history of medicine, Milstein, still haunted by the sorry tale, implied that it was the lack of sophistication of the NRDC and its patent attorneys that led to the unfortunate result. He thought that the NRDC people did not understand that one could patent an idea.[15] What he meant is that you could file for patents even without demonstrating that monoclonal antibodies had been used in diagnostics or therapy. To a large extent, even then, the patent laws of the world allowed for patenting prophetic descriptions of the "practical applications" that seemed to escape the blank minds of the NRDC functionaries.

No one is entirely free from blame in the Milstein matter. The NRDC had a failure of imagination. It did not seek out cutting-edge legal advice, even if it meant calling Rowland in California. Milstein waited to submit his *Nature* manuscript until there were only three weeks left to publication, too short a time to carefully evaluate how the patent law would deal with the new monoclonal antibodies.

But it was more than that: Milstein still had one foot in the precommercial biology of the early '70s. Without Milstein realizing it, the world was rapidly changing under his feet. He was being asked to place his other foot in the commercial world, and that did not come naturally to him. His spirit, like that of Paul Berg or of the early Stan Cohen at Stanford, was to do science for the public good, not for profit.

It was an either/or proposition for these biologists. There was something untoward about patents, and they thought that profit-seeking motives excluded the public good. If they wanted the public good, then no patents. If they filed for patents, they would somehow undermine the public good. It was a shortsighted and naive view of patents, and one that would change in the next decade. In chapter 4 I will explore the topic of patents in academia in more detail. I discuss there the difference between the fundamental technology shifts (such as genetic engineering or monoclonal antibodies) that we have been discussing here, and more routine follow-on applications of such shifts. I will explain my view as to why patents in academia serve to promote and raise necessary financial investments in risky life-saving technologies that may be follow-on applications of fundamental shifts in technology.

Figure 3.1 nicely illustrates the crossroads in which many biologists found themselves in the 1970s.[16] Some scientists—say, Milstein or Berg—are racing to the right on a track that curves left to the Nobel Prize, while others—say, Cohen or Boyer—are racing left on a track that curves toward wealth. In this discussion, "the Nobel Prize" will stand for scientific research with no clear commercial intent, done for curiosity or to save lives.

One scientist, scratching his head, seems lost at the intersection.

I think of the head-scratcher as a *she* and suggest that we call her Jennifer Doudna or Emmanuelle Charpentier. These two women scientists won the 2020 Nobel Prize in chemistry for their invention of CRISPR, a revolutionary gene-editing technology about which I will have more to say in the next chapter. Doudna and Charpentier have also been, and are still at this writing, involved in a fierce legal battle for the patent rights to their invention.

But the puzzled scientist of figure 3.1 need not be confused. As Doudna and Charpentier demonstrate, it is possible to get patents *and* win the Nobel Prize. One does not exclude the other. It is all a question of timing. If patent application filings are made before publication, then the scientists can have it both ways. Their dreams of tenure, fueled by the "publish or perish" ethos of academia and their wish to be first in the literature to compete for a Nobel Prize, need not suffer because of patents. They and their university can still get worldwide patent rights on commercial products coming out of their research if they time the patent filings ahead of the publications.

It is not a choice between patents or the public good. It can be both. Indeed, publishing without patenting deters investment in risky inventions. When discoveries and inventions end up solely in the pages of journals and do not benefit from further financing, few if any pharmaceuticals are made, tested, approved, and sold.

Stanford lucked out on patent filings because of Reimers, who by sheer serendipity in early 1974 fell upon the *New York Times* article on recombinant DNA. A few more months and there would have been no Cohen-Boyer patents, just like there were no Milstein-Köhler patents.

Yet out of these hesitant and ambivalent beginnings was born the biotech industrial-academic complex of today.

Figure 3.1. The choice of racing for the Nobel Prize or for patents. (Figure from Arti K. Rai, and Robert Cook-Deegan, "Racing for Academic Glory and Patents: Lessons from CRISPR," *Science* 358, no. 6365 [November 2017]: 874–76.) Reproduced by permission from AAAS.

## Gold Rush

The decade of the 1980s created a biological gold rush and changed the biologists' perceptions of their role in the world. It also opened the eyes of universities to the potential biowealth in their midst. Many universities in and out of the United States, watching Stanford's and UC's financial success with the Cohen-Boyer patents, started taking biological intellectual property seriously.

The beginning of the new decade saw the enactment of a federal law, the Bayh-Dole Act, which allowed universities to almost automatically own and exploit patents on inventions made with federal funds.[17] This move was not free from controversy. Critics of the law objected to patenting research findings after the work had been financed with taxpayer money. "Why should taxpayers pay twice?" asked the critics.[18] Yet, it was the congressional conclusion that university patents were necessary to promote industrial investment and bring downstream products to the market. Without such patents, private firms would be less inclined to make the necessary investments, and federally funded inventions would remain in the pages of science journals and never see the light of the marketplace.

The mining for biowealth in academia led to a few other developments. Niels Reimer's Office of Sponsored Research at Stanford, like many others, was renamed the Office of Technology Licensing (OTL). In the mid-1970s, universities created an association called the Society of University Patent Administrators (now known as the Association of University Technology Managers, or AUTM), to facilitate tech transfer from academia to industry. As of 2021 AUTM claims 3,100 members in sixty-five countries, including 800 universities. More than 200 university-invented drugs have been commercialized through the licensing mechanisms of the Bayh-Dole Act.[19]

Gone is the academic ambivalence of the early 1970s. It is now common for university patent administrators to pitch biological inventions to industry, oftentimes publishing magazines full of glossy pictures of lab-coated scientists peering at the ceiling through a test tube, apparently admiring the biotechnologies that will make them rich and maybe win them the Nobel. The OTLs negotiate directly with industry for research funding, including money for new buildings. The research in those buildings is then done with nothing but corporate money.

Today's venture capitalists show up on campus looking for inventions that might give rise to new biotech companies. A professor may disclose her government-funded invention of a new biotech drug to the OTL, help her university file for patents, and sign over her rights to the university. Then, with the help of venture capital money, she may spin off a company that, under the Bayh-Dole Act, takes exclusive rights to the university-owned patents. In return, her company pays royalties to the university. Sometimes, coming full circle, instead of taking money in the form of royalty payments, the university will take shares in the professor's spinoff. The university then becomes an investor in the professor's company and partakes in the growth of commercial development.

If things click just right—and they don't always—the public gets the new biotech drug, and everyone—the university, the investors, the spinoff, and the professor—makes a lot of money. Fifty years after Paul

Berg did the first genetic engineering experiments at Stanford, we all now live in the world of monetized biology.

## Atonement

It was the hesitancy of Old World thinking that led to the lack of patents for Milstein and Köhler's invention of monoclonal antibodies at the MRC. Others in Europe, however, soon recognized that they had to come to this side of the Atlantic to reap the benefits of our unique American culture of risk taking. The Germans were at the vanguard of traveling entrepreneurs searching for opportunities in this brave new world. And the Germans figure prominently in this story.

In 1983, just out of law school, and already in my law firm in Washington, DC, I was retained by Massachusetts General Hospital (MGH) in Boston, one of the teaching hospitals of the Harvard Medical School, to be its biotech patent attorney. Boston did not lack good patent attorneys, but in those early days of the Bayh-Dole Act and the academic-industrial gold rush, few understood biological science as well as those of us who had recently received our PhD in the field. My science degree from Harvard in 1976, my postdoctoral work at MIT in 1978, my law degree in 1983, and my unending excitement for the new biotechnology were my credentials.

It was a unique moment. Science know-how seemed to be more important to clients than gray hairs in patent law. Of course, it was also a good example of Harvard's old-boy network, but I didn't complain. I wanted to be the East Coast's equivalent to Bert Rowland, the imaginative California lawyer who represented UC and Stanford in the case of the Cohen-Boyer patents.

The German company Hoechst AG had, in 1981, signed an agreement to fund a new Department of Molecular Biology at Mass General.[20] It was to give MGH $70 million over ten years and have access to any intellectual property that came out of the collaboration. The hospital would own inventions and give Hoechst a first right to refuse a license. If Hoechst exercised its option, it would pay for the patents and get an exclusive license. I, as MGH's lawyer, was to collect invention disclosures, and propose them to the hospital's patent committee. If the committee approved, I would then write and file patent applications. The committee was headed by a patrician, gray-haired senior partner from Ropes and Gray, the venerable Boston law firm. He liked me, but he couldn't fathom what this young whippersnapper from out of town was doing on his team.

Once a month I would fly up from DC to Logan Airport on the early Eastern Airlines shuttle and return in the evening of the same day. I would arrive at Mass General at ten o'clock in the morning and meet with Marvin Guthrie, who was the equivalent of Stanford's Niels Reimers at MGH. Guthrie was one of the founders of the Society of University Patent Administrators and one of its first presidents. Together, Guthrie and I would visit scientists. Our charge was to wander up and down through the labs of the esteemed MGH and harvest IP. The scientists talked to me because they saw me as one of them, never mind my law degree. Maybe I was a scientist who had gone off the beaten track of academia onto a road less traveled, but a scientist nevertheless.

The head of the "Hoechst Department" was Professor Howard Goodman, one of the central figures of molecular biology at the time. Indeed, it was Goodman who had initially gone to Hoechst and proposed the whole collaboration. One of the beauties for him and his scientists was the basic nature of the research they could do, with no pressure from Hoechst to focus on commercial work. If any IP came out of it, the better for Hoechst, but at least in all public comments and press releases, that was not the main goal of the agreement. Plus, Goodman and his colleagues did not have to apply for research grants anymore, a burden well known to all federally funded scientists. Goodman convinced a constellation of biological stars to come along. Among them were later luminaries, such as Jack Szostak, the 2009 Nobel Prize winner for his work on chromosomes; Fred Ausubel, a worldwide expert on the biology of nitrogen fixation in plants; and Brian Seed, a pioneer in immunology. The whole thing was a "showstopper," as Barbara Culliton called it in her 1982 article in *Science*, "The Hoechst Department at MGH."[21]

In 1984, as part of the agreement with MGH, Hoechst completed the remodeling of new laboratories for the department, known as the Wellman

Building. I was at the ribbon cutting that morning. It was attended by the who's who of Harvard's biological science and its sponsors. The chairman of Hoechst flew over from Germany and led the ceremony. I expected a standard speech praising the virtues of academic-industry collaboration, the brilliance of the folks in the Wellman lobby, and American-German friendship. It was all there.

But there was more, and his extra comments gave me an unexpected jolt. The chairman came right out and said that this moment was important not only because of all of the above but because it symbolized how far Germany had come since the 1930s and '40s, when it had either expelled or later vanquished its Jewish scientists. Hanging in the air was the knowledge that Hoechst AG had been one of six German companies, which were part of IG Farben, a major chemical conglomerate during the Second World War. IG Farben had supplied some of the components of the deadly Zyklon B gas, used to murder Jews and others at Auschwitz. After the war, IG Farben was broken up, and Hoechst AG was reborn as an independent company.

German science had never recovered from the loss of its Jews, added the chairman in perfect although slightly accented English. German biotechnology was lagging the United States, and the only way to catch up was to come to the US ". . . and learn from you" (or something to that effect). And, although he did not say it, we all knew that Howard Goodman, the head of the new department, was a Jew and that the labs had plenty of other Jewish scientists. The speech acknowledged two distinct realities: one, the profound loss of German research talent and, the other, the recognition that a large part of the cutting edge in commercial biology was here in the United States.

Contemporary Germans, more than anyone else, have learned how to atone for the terrible things their ancestors did in the Second World War. They have also gained a great deal of modesty since the days when they thought themselves masters of the world. The inauguration of the MGH Department of Molecular Biology was an event that gave Hoechst AG a chance for a moment of redemption. The honest and moving ribbon-cutting ceremony in Boston has remained with me over the decades.

# 4

# Why Patents in Academia?

At this point you might say, I understand the basic concept of patents: society wants to incentivize private companies that have a penchant for secrecy to disclose their technologies, so we let them have exclusivity. But what about universities like Stanford or UC, whose professors Stan Cohen and Herb Boyer invented recombinant DNA technology in the mid-1970s? And what about the MRC in England, whose scientists Milstein and Köhler invented how to make monoclonal antibodies? These four scientists needed no incentive whatsoever to invent and quickly disclose their pioneering technologies. To them, patents were afterthoughts, not motivations. Why, then, do we even need patents in academia?

## Good Questions

As we saw in chapter 3, it was Niels Reimers, of the Technology Licensing Office at Stanford, who read in the *New York Times* that something momentous had been invented in the biolabs. Ironically, he only found out about this breakthrough because it had *already* been published. It was only at that moment that the university woke up and filed for US patents. Across the Atlantic, and with a few weeks left before their paper appeared in *Nature*, Milstein and Köhler sent information about it to the NRDC, the entity in charge of patenting inventions. We know how that ended: the NRDC ignored him and obtained no patents at all.

Like all academic researchers of the time, these four biologists published because that was their professional goal: to allay curiosity, to help humanity, to disseminate knowledge, to get tenure, to become well known, maybe even to receive a Nobel Prize. Keeping secrets was not an issue. Quite the contrary: there was no need to give them an incentive to invent and then publish their work. They did so spontaneously, with pride and gusto. Our four molecular biologists willingly told everyone about their "ingenious contrivances," as the Venetians called them. They did not need a patent to do so. And the lack of strong patent protection did not inhibit investments in the university-invented recombinant DNA and monoclonal antibody technologies. Both fields exploded, even in the absence of patent exclusivity.

It may therefore look as if university discoveries are the furthest from the deal the Venetians had in mind in 1474: a patent for your secrets. So, does the basic Venetian bargain of exclusivity for disclosure fail for university inventions? The answer is, not quite.

## Technology Shifts and Applications

The big picture of how an idea originates in the mind of a professor, is then proven in her lab, and, if commercially interesting, ends up in a marketable pharmaceutical is more involved than a simple binary equation: whether to publish or to patent. In the case of average university inventions, the equation is not about disclosure; *it is about investment*. It is not exclusivity in exchange for publication; it is about the other side of the patent incentive: exclusivity in exchange for private financing. The patent system exists primarily to diminish the risks of investing in the further development of inventions. Investors hate risks, and patents ameliorate risks.

In chapter 3, I (slightly) misled the reader by using recombinant DNA and monoclonal antibodies as illustrations of average university inventions. These two examples, however, were, and remain, far-from-average innovations. They were what I would call *technology shifts*. Technology shifts are fundamental innovations that, once invented, are used by everyone because there is no longer any choice but to do so. The previous ways of doing things are displaced, perhaps forever. Shifts also open new fields where there were none before. Think about the invention of the printing press and how it eliminated hand-copying by monks. Or the transistor, and how it replaced the vacuum tube; or the handheld computer, and how it banished slide rules to the museum. Why, after these technology shifts, would anyone copy books by hand, or use vacuum tubes or slide rules? Very few would.

Now consider how genetic engineering is used to produce human insulin in bacteria, and how it replaced the previous labor-intensive extraction and purification of insulin from pigs. Once purified, the pig insulin had to be chemically modified so it became more like the human variant. Why would anyone use pigs to extract and modify its insulin when they can use genetic engineering and make human insulin directly in bacteria? Again, very few would.

Patents are not needed for investment into technology shifts. When a university lab is lucky enough to have generated a tech shift, such as recombinant DNA technology or monoclonal antibodies, there is no need to incentivize anyone to start using it to create commercial products. Quite the contrary: everyone rushes to use it. The invention and patenting of a shift are together a windfall for academia.

But technology shifts are rare. Most university inventions add much less to the sum of knowledge than do fundamental upheavals. Most innovations, whether in academia or industry, are not as transformative but instead take advantage of existing technology to improve on the way things were done in the past. Once genetic engineering was invented, it was promptly used for making human insulin, pest-resistant cotton, and new biologic drugs, like the anti-rheumatism drug Humira. These are just three of many biotech inventions that followed the biological upheaval of the mid-1970s. And while better insulin, insect-resistant cotton, or anti-rheumatism drugs are dramatic inventions, they are what I would call, without wishing to sound dismissive, *applications* of the new breakthroughs. Such applications still require a period of limited exclusivity that gives private investors some breathing room before being copied. Publishing without filing for patents denies the possibility of exclusivity, and such applicative inventions languish without the investment necessary for commercial development.[1]

Should society then propose that universities not be allowed to receive patents on technology shifts but only on their applications? While such a question may be worthy of a good debate over a beer late at night in a law school dorm, it opens the door to endless argument and hairsplitting. Who, in the early days of a university innovation, is to decide whether it is going to be a technology shift or a stepwise advance? Universities? Companies? Both will insist that patents are needed to provide incentives for investment. The government? That would be the worst idea of all. A judgment by a government agency, coming early after the appearance of a recent invention, is fraught with uncertainty and political manipulation. It would lead to never-ending litigation about what is a shift and what is an application.

I am not an unskeptical believer that the market always knows best, but on balance I would leave things as they are. As we saw in chapter 3, that was the judgment of the US Congress in enacting the Bayh-Dole Act in 1980, allowing universities to own and patent *all* inventions made with public funds.

Every now and then, university research will lead to a technology shift. If so, I believe that we should allow the university to patent it and benefit financially. We should at least make sure that federally funded tech shifts are not licensed exclusively to any one company. They should be made available to all, so that, as was the case with the patented invention of recombinant DNA by Cohen and Boyer, a whole new industry is created. The new industry can then use the tech shift to generate additional inventions for the public good. The occasional university patent on a technology shift is worth the price of a windfall. Likely, the university will invest the windfall in more research and investigation. In any event, such a result is good for society.

And because there is no legal difference between patents on technology shifts or applications, or between inventions made in academia or at private companies, the battlefield over patent rights in commercial biology is populated by both universities and businesses. Sometimes companies litigate against each other, sometimes they litigate against universities, and sometimes universities litigate against each other. Indeed, less than a decade after the 1970s, I started seeing professors fight each other fiercely over patent rights. This was an extraordinary development compared with the old days, when laid-back biology departments had nary a thought of patents. I remain startled to see how far things have changed since the early years.

Let me tell you of two university inventions with which I was—and still am—closely involved. One was an application, the other a technology shift. The first was a fight over who invented the anti-rheumatism drug Enbrel®. Part of the Enbrel invention came out of the Department of Molecular Biology at Mass General. Enbrel was an *application* of basic genetic engineering. The second story is the lengthy—and still ongoing—litigation over who first invented CRISPR, a genetic technique used to "edit" DNA; that is, to exquisitely change one or more "letters" of the genetic code. CRISPR came out of multiple universities and is a technology shift. Both the invention of Enbrel and that of CRISPR caused multiple companies to have spent and continue spending fortunes to own and exploit them.

## The Enbrel Jackpot

As I described in chapter 3, in 1981 the German firm Hoechst AG committed $70 million to fund research at MGH's new Department of Molecular Biology. The Germans were publicly interested in learning recombinant genetics from American scientists. But, of course, they were also interested in commercially useful inventions and discoveries. And they were interested in patents.

During monthly visits to Boston, I would wander the labs in hopes of garnering patentable inventions. A German patent attorney, whom I will call Dr. Manfred Gross, would fly over from Frankfurt every now and then and stay in Boston for a few days. He would accompany me on such visits. Gross and I became instant colleagues, both of us excited by the whole concept of patents in the new science. He was older than me, yet, because of my recent training in molecular biology, deferred to me when it came to discussions with MGH's potential inventors.

A memorable conversation took place in the mid-1980s with the head of one of the MGH labs. It started as a modest scientific discussion and ended up anticipating by several years one of the most successful drugs ever invented for the treatment of rheumatoid arthritis. The story has all the elements of an invention made in the age of commercial biology. It began in one academic institution, quickly became entangled with discoveries made by scientists at other universities and at different companies, and led to massive litigations both at the inception and at the end of the drug's patent life, peppered by multiple corporate acquisitions.

The lab head welcomed Gross and me into his small office with a friendly smile and went right to business. He walked over to the whiteboard and drew a schematic of what looked like an antibody molecule. "You know what an antibody looks like, right?" he continued. We all nodded. I thought, "Well, what I think of as an antibody and what he thinks of as an antibody may not be identical, but I will not interrupt." So I, too, nodded.

An antibody is a Y-shaped molecule. See the lower left of figure 4.1. The four diagonal ends at the top of the Y are the portions where the antibody binds to foreign invaders like viruses, so the body can defend against them. The lower, vertical two portions, are a "stem" that binds to cells of the immune system and mediates the body's defensive effects. "We have been working on antibody fusions," said the lab chief. "These are attachments of portions of two different protein molecules that are not joined to each other in nature. One portion is the stem of an antibody," he added, completing the drawing.

The second portion of the fusion was that of a "cellular receptor." The fusion he drew used only the vertical stem portion of the antibody and replaced each of the four diagonal tops with a piece of such a receptor, specifically the receptor for Tumor Necrosis Factor (TNF). TNF is a highly inflammatory molecule. In nature, when TNF streams along in the bloodstream, its receptor, which is anchored to the surface of a cell, captures it and sends a signal to the inside of the cell. The signal then triggers inflammation, such as arthritis. The static receptor is like a high school proctor sitting in front of the principal's office watching out for

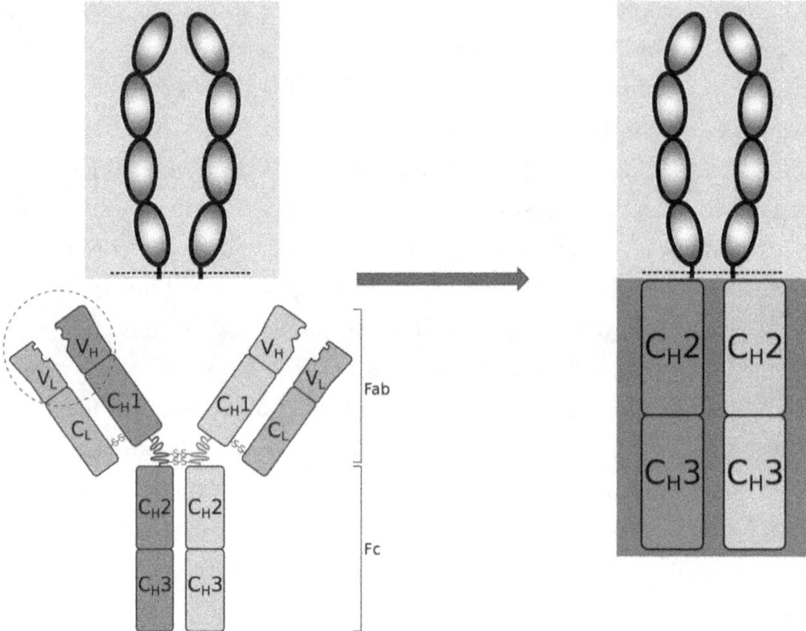


Figure 4.1. A schematic showing the construction of Enbrel. *Above left*, a portion of the TNF receptor. Source Wikimedia Commons marked copyleft; attribution: NEUROtiker. *Lower left side*, whole antibody. Source Wikimedia Commons, licensed under the Creative Commons Attribution-Share Alike 4.0 International license; attribution TOKENZERO. *Right side*, Enbrel, as a fusion of the lower part of the whole antibody and the portion of the TNF receptor, assembled by the author.

troublemakers. When one happens to come by, the guard catches him and triggers a signal by calling the heartless headmaster, the one whose name is "Dr. Arthritis."

In the antibody-receptor fusions we were discussing that day, however, the receptor is not affixed to the surface of a cell anymore. Instead of sitting at his office door, our proctor is now freely wandering and mingling in the school hallways. There, he keeps watch for troublemakers that are moving along. Just like the proctor waiting at the door, when the circulating TNF receptor runs into the inflammatory TNF, the circulating proctor binds to it and takes it out of commission before it causes further trouble. As a result: no call to the heartless headmaster, no arthritis. The idea was that creating a freely circulating fusion molecule with a TNF receptor on it would soak up TNF floating around and, as a consequence, decrease arthritis. Because the receptor would catch TNF before it bound to a cell, it was called a "decoy." This decoy was a close predecessor of Enbrel.

Figure 4.1 shows a schematic on how Enbrel is put together. First, on the lower left of the figure, is the whole antibody with three arms: two diagonals and one vertical. The biologist separates the two diagonal arms from the vertical stem and saves the stem. The important portions of the TNF receptor are the two strings of four beads each shown at the upper left of figure 4.1. These two bead strings get joined to the saved vertical portion of the antibody. The resulting fusion, which is shown on the right side of the figure, has one portion from the antibody and another portion from the TNF receptor. This is Enbrel.

The joining of the different portions would be done through the DNA encoding of each one, explained the MGH scientist. The fused DNA would be transferred into a bacterium, and its machinery would produce large quantities of the artificial protein, which could then be isolated and purified. Think of the DNA for an antibody fusion in pop bead terms. A first string of a hundred pop beads, representing the vertical antibody stem portion, would be blue. A second string of a hundred pop beads, representing the four-beaded TNF receptor, would be red. Now, click the end stub of the blue string to the end stub of the red one, and you have made a lengthy bead string with two colored sections, each corresponding to a portion of the fusion molecule. I think that most recombinant DNA geneticists must have played with pop beads in their childhood. Many of the ones I have met in my life took the same delight in recombining genes as a child mixing and matching pop beads. That's what the scientists in MGH's research group were doing with strings of DNA.

When I heard of the TNF-antibody fusion that morning, I did not know then that MGH had hit the jackpot. I also couldn't know how much legal trouble would be created by multiple disputes about the patent rights to this one ingenious molecule.

Shortly after MGH's patent filing, our legal team discovered that there were at least three other groups around the world that were thinking of the same thing: there were scientists at the Seattle company Immunex; at Hoffman-LaRoche, in Switzerland; and at a research group at the University of Texas. All of them filed their own patent applications, and a priority race was on. Pretty soon the landscape for who would get the patent rights to the TNF decoy fusion became as muddied as a super-truck jamboree on a rainy day.

To understand the priority mess over Enbrel, we need to go back to the days before 2013, when United States patent law changed from what was known as a "first to invent" system, to a "first to file" system. We have never liked simple legal solutions in this country, especially when most of the rest of the world asks us to please do what it does. The simplest solution as to who gets the patent in a four-way race would be to see who filed first. End of report.

But no, for decades, until the United States finally joined the rest of the globe, our law gave the patent to whoever invented first, not necessarily who filed first. This exceptional American way of doing things resulted in a complex legal procedure called "an interference." The name reflects the fact that multiple pending patent applications to the same invention devised within a short period of time by different inventors "interfere" with each other. Each inventor would file evidence of his conception and his reduction to practice. That is, on what date did the light bulb go on? And, on what date was the idea shown to work in the lab? This area of patent law was, to put it mildly, abstruse.

I spent many years of my career as an interference specialist. I consumed endless days straining my eyes, trying to understand the barely legible notebook entries of scientists whose hieroglyphic scribbles only they could decipher. I hoped against hard-won experience that they had dated their pages, as that was the most important piece of information: When did this or that experiment happen? And what did it show?

Since 2013 the United States no longer has interferences and, much to the delight of the rest of the world, has a simpler, "first to file" system.

You look at the filing dates, and you know who wins. No need for expensive lawyers consulting Rosetta Stones.

By the 1980s and 1990s, however, when MGH's decoy invention was at stake, interferences were the law of the land. My client was facing a complex, four-way interference. When all of us contestants looked at the muddy landscape on who would win priority, we realized how expensive and lengthy a battle this would be. After much legal wrangling and negotiation, Immunex acquired everyone's patent rights, and the priority disputes ended. The question of who was the first to invent was resolved in a more informal and less expensive way than multiyear litigation. Immunex, like an oil prospector who buys acreage from different sellers without knowing for sure where the lucky spot will be, would have rights to all patents and would consequently "win" the priority. The winner was Professor Bruce Beutler of the University of Texas.[2] His patent issued in 1995. I will have a lot more to tell you in chapter 19 about the litigations on Enbrel that took place late in the life of Beutler's patent, around 2012. Meanwhile, let me continue with the story of Enbrel's early life.

When Enbrel appeared in 1998, more than a decade after the initial patent filings, the drug became an instant pharmaceutical game changer. Dr. David Felson, a professor of medicine at Boston University Medical School, and one of the world's experts on rheumatoid arthritis (RA), recently said to me, "After Enbrel, the problem of RA has been solved." Gleaning the worldwide prevalence of RA in an aging population, and the riches to be made with such a blockbuster drug, in 2002 Amgen bought Immunex and its rights. It also bought Immunex's obligations, such as having to pay the MGH royalties on worldwide sales of Enbrel. In 2006–07, MGH settled its rights to receive royalties from Amgen for about $470 million.[3] Not bad for playing with pop beads.

The mid-1970 inventions of recombinant DNA or monoclonal antibodies made in academia were technological shifts. Whether they were protected by patents or not, everyone adopted them, and they became platforms for the myriad applications that followed. These shifts were so fundamental that no one wanted to go back to the old ways of doing things.

The invention of Enbrel, however, was an application of the shifts. It was a spectacular drug that changed the way rheumatic conditions would be treated going forward. And, as evidenced by the interference litigations and the many corporate acquisitions that followed, Enbrel needed to be protected by patents before it could be made available to the public.

Whether Mass General intended it or not, the hospital had a windfall because of the then recent phenomenon of biological monetization. I am certain that the academic scientists were not incentivized to invent in exchange for potential financial gain. They told me so. At the end of the day, however, they received a share of the profits. So did MGH, which plowed most of them back into research funding. The hospital, with its brand-new Department of Molecular Biology, was the right academic institution, at the right time, at the right place.

## The CRISPR Battles

It is illustrative to compare the early patent battles for Enbrel with those for CRISPR. CRISPR is a technique that allows flawed genes to be made flawless. If a live entity such as a microbe, plant, or animal has a faulty gene—say, one with a mutation that may be debilitating or even fatal—CRISPR editing can be applied to make the imperfect gene perfect and restore the entity to a healthy life. And if the entity is a human being, CRISPR can be used to cure inborn genetic defects, such as Huntington's disease, sickle cell anemia, or high cholesterol.

The technique is one of the most fought-over academic inventions of all times. It has led many and prestigious universities to become aggressive legal adversaries to determine who first invented it. The litigation over rights to CRISPR is, as was the fight over Enbrel, a patent interference; actually, as of this writing, it is multiple patent interferences. The CRISPR interferences are historically among the last ones that the US patent system will ever see. The procedure of determining who was first to invent was removed from our patent laws in 2013. Since the CRISPR inventions go back to 2012, they went into interferences by the skin of their teeth.

Our law firm represents one team in the interference fights. The team includes the University of California, home to one of the inventors, Professor Jennifer Doudna. Also on the team is her coinventor Professor Emmanuelle Charpentier, as well as Charpentier's previous home, the University of Vienna. We call our client team "CVC": California-Vienna-Charpentier. Charpentier, who at this writing is the director of the Max Planck Unit for the Science of Pathogens in Berlin, received with Doudna a joint Nobel Prize in chemistry in 2020 for their invention of CRISPR.

The CVC scientific team is fighting tooth and nail over valuable patent rights to CRISPR against another team headed by Professor Feng Zhang of Harvard University, Massachusetts Institute of Technology, and Broad Institute—all in Cambridge, Massachusetts. All the academic institutions on "both sides of the *v.*," as we lawyers say, have major private financial backers who stand to benefit immensely if their side wins exclusivity.

The science behind CRISPR is as complicated as it gets. Since you are no doubt curious, I will tell you: CRISPR stands for "Clustered Regularly Interspaced Short Palindromic Repeats." (Don't worry, there will be no test at the end.) The tongue twister describes a genetic structure ubiquitous in bacteria that has evolved over the eons. It allows bacteria to recognize a virus trying to infect them, and if they have encountered the virus before, to reject it. It is, in other words, a primitive immune system. Doudna and Charpentier invented a modification of the natural system, which allows molecular biologists to accurately change one or multiple letters (A, C, T, or G) in the four-alphabet code of DNA to correct genetic sequences that are incorrect. Zhang and his team, however, say that they invented it first; or, at least, that they were the first to make it work in mammalian and other complex cells. That is what the CRISPR interference is all about: who conceived first, in what, and can they prove it?

To give you an idea of how profound an ability it is to change one genetic letter at will, you need to know that the human genome has 3 billion letters. If we equate each letter to an inch, then 3 billion inches is roughly twice the length of the earth's equator. CRISPR allows you, in an exquisitely elegant manner, to rapidly find one of those letters and exchange it—and its neighbors too, if you so wish—by another one. Imagine being tasked with quickly finding a unique one-inch-long letter T somewhere along the length of two equators and replacing it with a one-inch-long letter A. Since about one-fourth of the letters in the genome are Ts, there are about 700 million Ts placed randomly along the two equators. Using CRISPR, you can readily find the unique T and exchange it for the A. And if the T you are looking for is the single point mutation that causes sickle cell anemia, you can clean up the genetic error and cure the patient.[4] In 2021 the FDA approved a clinical trial using CRISPR to do precisely that.[5]

As of this writing, the battle for priority of the CRISPR invention is far from over. In 2021 it was expanded to include two additional companies: ToolGen of Seoul, South Korea, and Sigma-Aldrich of St. Louis, Missouri. Consequently, it is now a free-for-all legal war, with six distinct interferences in the United States alone. If you are inclined and wish to learn more about CRISPR, you may read Walter Isaacson's excellent book, *The Code Breaker*.[6]

CRISPR is one of those biological innovations worthy of being called a technology shift. The invention happened around 2011–12, thirty years after the first recombinant DNA molecules were made by Paul Berg. While the invention of Enbrel in the 1980s may be characterized as an application of the then still young science of genetic engineering and CRISPR's invention in the 2010s is more of a shift, they have a few things in common.

One is that they led to interferences with multiple parties fighting each other for the patent rights. Another is that many of those parties were and are universities. And in both instances, companies quickly understood the commercial value of both inventions, took licenses from the universities, and spent serious money to control the ultimate patent rights. Intense interacademic legal battles as CRISPR were an inconceivable proposition in 1971, when Paul Berg at Stanford made the first recombinant DNA molecule. Berg did not even think about filing patent applications. He belonged to an earlier, less commercial biology world. The commercialization of biology changed his world.

I cannot help but think that had CRISPR been invented in the decade of the 1970s, as was the case with recombinant DNA or monoclonal antibodies, no or only a limited set of patent applications would have been filed. After publication, everyone would have adopted the technique, just as everyone adopted recombinant DNA or monoclonal antibodies. There would not have been any fights to try and figure out who will own the patents.

But 2023 is not 1971. Fifty years have passed, and much of the academic world of biology is now focused on business.

# 5

# Scientist-Lawyers

The rise of commercial biology had an impact on many realms, whether in patenting, financing, or academia. And it also had a profound influence on the law profession. It produced a new set of legal "hybrids." As so many of the other hybrids that were created by the new biology—Berg's chimeric viruses, Cohen-Boyer's recombinant DNA molecules, and Genentech's entrepreneur-scientists—we in the law world also generated hybrids.

My partner Eldora Ellison is one of them: a hybrid of law and science. And the increased role played in biotech IP by women lawyers such as Ellison is also a wonderful upshot of the biological developments of the last fifty years.

## Women Patent Attorneys

Let's go back to CRISPR, the gene-editing technology we discussed in chapter 4. Understanding the Clustered Regularly Interspaced Short Palindromic Repeats well enough to explain them to a judge is not a simple proposition. Eldora Ellison, as the lead counsel representing the California-Vienna-Charpentier (CVC) team, oversees the understanding and explaining. Knowledgeable, unruffled, with a good sense of irony and humor, Ellison is the ultimate scientist-lawyer. Figure 5.1 is a photo of Ellison in her office, around the time of the CRISPR interference.

It is fitting that Ellison, a 1993 PhD in molecular biology from Cornell University and a 1999 JD from Georgetown University Law School, represents Professors Doudna and Charpentier, the two Nobel Prize

Figure 5.1. Eldora Ellison, PhD, JD, in her law office, ca. 2020. Reproduced by permission from Sterne, Kessler, Goldstein & Fox PLLC.

winners, in their legal battle against the Zhang team. It is gratifying to see two famous women inventors represented by a woman patent attorney in the mother of all biotech patent fights. As the scientist-lawyer that she is, Ellison commands the biochemical mechanisms of CRISPR and the law of patent interferences like no one else. But it is not only in science and law that Ellison excels.

One of the best recollections I have of Ellison's unflappable legal temperament is of her presiding over a large team of lawyers in the CRISPR case. When all the interested parties on the CVC side met in our offices, there were normally between thirty and forty lawyers in the conference room. They included counsel for the University of California; the University of Vienna; for Professor Charpentier herself; for Charpentier's company CRISPR Therapeutics, Inc.; for the license holders of the University of California; and even for the sublicense holders of the license holders. You cannot place so many restless legal minds in one room for a day without a fair number of intense opinions and speeches. Sooner or later, somebody will have something to say, even if lengthy and repetitive.

Ellison had the habit of sitting at the front of the table and starting the meeting at 9:00 AM sharp. She first welcomed everyone and then, for the benefit of those who were not present but were there by video, quickly recited the names of all participants, live or virtual, one by one. *And she did it from memory*. There was absolute silence in the room, as we all expected that she would draw a blank. But she never did. Her performance was flawless, and the room often broke into applause at the end. After that tour de force, she had command of the lawyers for the rest of the day.

Ahead of the 1980s—when inventions were focused on machines, circuits, drugs, and plastics—patent attorneys came from chemistry, physics, and engineering. Before biology became the subject of patents, the traditional technical careers required by the Patent Office to pass the patent bar were devoid of women; most patent attorneys tended to be male. Consequently, the number of women in the patent profession had always been low. By late 2012 only about 18 percent of overall patent practitioners were female.[1]

The biotechnology revolution changed that. There are more women in the United States who study biology than electrical or mechanical engineering. And with more women biologists, more of them go to law school and become patent attorneys than before. The ratio of women to men in our practice, and probably in that of most other biotechnology law firm practices in the United States, is close to 50:50. This ratio is even better than the 46 percent of women in the general legal profession.[2] The rise of commercial biology has been good for women patent attorneys such as Ellison.

## Flying the Coop

Preparing patent applications and procuring patents in cutting-edge technological areas has always required a deep knowledge of both science and law. The biotech intellectual property world is no exception. While the trend to hire scientist-lawyers who understand biology and law accelerated in the age of commercial biology, scientist-lawyers have been around for many years. Most have come from scientific careers that have bored them or left them dissatisfied.

Alan Lourie and his colleague Pauline Newman, both judges at the Court of Appeals for the Federal Circuit, are good examples of scientist-lawyers. Judge Lourie was a Harvard undergrad, received a PhD in organic chemistry from Penn, became a chemist at Monsanto, then a patent agent at Wyeth (the pharma company known for Advil®), and an in-house patent counsel at Smith Kline Beecham (known for Sucrets® and Tums®). He became a federal judge in 1990. Judge Newman's background took her from Vassar, to Columbia, and to Yale, where she received her PhD in chemistry. She became a federal judge in 1984. Both are scientists and legal scholars. They understand chemistry, and their decisions are notable for the clarity of their reasoning.

Reflecting on his own transition from science to law, Judge Lourie once said to me, "We flew the coop." My own move to law mirrored Lourie's and Newman's. The shift was not as difficult as I thought it would be. I quickly understood that the common law, as practiced in the United States through the analysis of appellate precedents, has a lot in common with the scientific method. The first relies on a study of court cases and the second on a study of data. I was used to studying data to see whether Mother Nature was revealing a predictive pattern. I saw the cases in my law books as the data of my research days. I studied law cases and recognized that the courts were also struggling to come up with some sort of pattern. I came to appreciate that the beauty of the case law system is that many of the "rules of law" come from the ingenuity of individual lawyers arguing their cases, not from some deus ex machina that hands down "the law" from high up. I enjoyed myself immensely studying law. Flying the coop was not as hard as I thought it might be.

The mid-1970s had seen two major technological revolutions: informatics and biotechnology. The rise of IP followed. Microsoft appeared in '75, Genentech in '76, Biogen in '78. The year 1980 saw the founding of Amgen and Genetics Institute, and Wall Street celebrated Apple's IPO, its initial public offering. All that these high-tech startups had in store was intellectual property, and IP swiftly became everyone's darling. It was the new gold, the product of advanced thinking, nothing less than the reflection of our national ingenuity. Our enemies desired it, they spied to find it, and they stole it when they saw it. The world of IP law changed, and patent attorneys seemed to be at the right place, at the right time, and in high demand.

I graduated from George Washington University law school in 1983, amid all this scientific and commercial ferment. I realized in law school that being able to bridge two worlds, one from science and the other from the up-and-coming biotech industry and its legal needs, would be a good thing to do.

The young biotech CEOs who were leading the Genentechs, Biogens, and Amgens of the world were looking to protect their intellectual property and, more than anything, were looking to find lawyers who understood the new biology. They wanted attorneys who could chat in the language of their research scientists. The new talk was all about "recombining DNA to produce human proteins in genetically engineered bacteria," "making monoclonal antibodies," or "using the latest promoters for expression in mammalian cells." This is complicated stuff, and experienced patent lawyers didn't easily convince the CEOs that they were able to manage it. Knowing the jargon is never the same as understanding what it all means. There was a palpable moment when it looked like grasping the new biotechnology was more important than years of legal experience. Gray hairs didn't seem to matter as much as recent scientific training.

## Hybrids on Both Sides

Jim Haley, a contemporary of mine, is another classic scientist-lawyer. He received a PhD in organic chemistry from Brandeis and a law degree from Suffolk, both in 1975. He climbed the ranks and became a partner at the prestigious patent law firm of Fish and Neave (F&N), in New York. F&N was the firm that in 1906 obtained for the Wright Brothers a patent on their "Flying Machine" and had been patent counsel to some of the most famous inventors of the times, such as Thomas Edison (the light bulb), Alexander Graham Bell (the telephone), and Edwin Land (Polaroid pictures). F&N was an establishment, a gold-plated firm of highly respected patent attorneys, referred to by others[3] (especially by its partners[4]) as "the princes of the patent bar."

Haley is a very creative biochemist and patent attorney. Among his most enduring contributions to patents in biology is the use of "DNA hybridization" as a way of defining the similarity between two DNA sequences in encoding the same protein. When two different DNA sequences are highly similar (in a complementary way), they will readily bind to each other and form a double helix; they will "hybridize" to each other. They are then more likely to encode the same protein. In contrast, when they are not highly complementary, they will not hybridize, and they are then less likely to encode the same protein. There are thousands of patents around the world that use Haley's idea of DNA hybridization to compare DNA sequence coding similarities. *Chapeau,* as they say in France.

I had several encounters with Haley. Perhaps the most memorable one was at an oral hearing before the European Patent Office in Munich in 1986. I remember it because it was one of the first public trials of my career. The litigation was over patent rights to the gene for alpha interferon, which recently had been isolated by Charles Weissman, a professor in Zürich. Weissman had granted ownership of the patent to Haley's client Biogen, of Geneva, Switzerland. I represented Boehringer Ingelheim (BI), a German pharma company that was one of nine parties opposing the recent issuance of Biogen's patent. Haley and I were at opposite sides of the case, and at far ends of a large circular table.

The modern hearing room looked less like a classic US courtroom and more like the United Nations Security Council. Haley, I, and many others sat with our clients around the vast table. Since there were ten litigants, the rows of lawyers and clients ran two to three deep. This was one of the earliest biotech litigations in Europe, and lots of curious legal types from around the world showed up just to see what would happen. The remarkable thing was that in the early 1980s there were very few, if any,

European patent attorneys with knowledge of biology. So Biogen flew Haley, and BI, its opponent, flew me, across the Atlantic to help. The creation of scientist-lawyers, "PhD-JDs" as we are labeled, had not yet started over there, so the two companies imported Americans for the hearing. There were hybrid lawyers "on both sides of the *v*."

The proceedings lasted a whole day. Everyone had a chance to say something, including Haley and me. The four judges listened carefully and took copious notes. The opposition trial ended by 5:00 PM, and the judges told us to come back in two hours. When we returned, they ruled against Biogen and revoked the patent. It was swift and sweet, and our side went out for a Bavarian victory meal with plenty of beer. Our success, though, lasted no more than three years. Haley appealed, and in 1989 the appellate division reversed and reinstated the patent.

Haley was a worthy opponent; when together we always had mutual respect and a glint in both our eyes. There is nothing better for a patent lawyer than to have an adversary who understands the science and the law. Both lawyers can cut to the chase quickly and not waste their time or their clients' money.

## Salamanca

Lest you conclude that my admiration for Judges Lourie and Newman, or for colleagues like Ellison and Haley, leads me to believe that all scientist-lawyers walk on water, let me disabuse you. Not all PhDs will turn into top legal talent like these four. In forty years of hiring and training researchers to become attorneys, I've come to realize that a PhD degree is no guarantee of intelligence.

A Latin phrase carved into an entrance stone at the eminent University of Salamanca in Spain cautions all comers: "Quod natura non dat, Salmantica non præstat." What nature does not give, Salamanca does not lend. In other words, all ye who come to these gates seeking a degree, be warned that if nature hasn't endowed you with smarts, a doctorate from Salamanca will not give them to you. Smarts are the ability to see what is not immediately obvious, to connect dots, to be creative. If our scientist doesn't have smarts, then sending him to law school will only produce a dim lawyer with a PhD.

But neither brilliance nor insight alone will make a good lawyer; these qualities are necessary but not sufficient. A top patent attorney needs to be a good advocate, relaxed when arguing a point of law with a judge and passionate when trying to persuade. She must be able to explain complicated scientific matters to a jury whose average technologic savvy may go no further than using a remote. He needs to be personable, with a good gift of the gab when talking to potential clients. A good lawyer needs to have a feel for the jugular and go for it. She needs to find her opponent's weakest spot and press hard. He needs to write in ways more eloquent than the woodenness of his old scientific papers. And importantly, she needs to stop being a scientist and become a lawyer.

"Yes," I say to my new associates, "clients may hire you because you speak their language and understand what's going on in their labs. But they don't hire you to be a scientist. They hire you to be a lawyer. Never forget that." I have seen many a brilliant molecular biologist fumble when in front of a panel of appellate judges or when trying to write a convincing legal brief.

Many scientists don't make the transition. If after a year in our firm I still see copies of the journal *Science* on their desks, I have that sinking feeling that they haven't been able to shed their old lab coats. I have found that reading *Science* for enjoyment is a tell-tale clue that they haven't moved on. They ought to be reading the latest decisions from the Court of Appeals, I tell them, or the commentary in the *AIPLA Journal*, the scholarly magazine of IP lawyers.

Sometimes, the habits of old scientists are hard to break, even for top legal minds. They appear in subtle and unexpected ways. None other than Judge Lourie, one of the top scientist-judges on the Court of Appeals, once had difficulty distinguishing science from law. Lourie's propensity to see legal issues through the prism of science got him in trouble with the Supreme Court, which reviews decisions of the court where he sits. As I will show you in chapter 10, in *AMP v. Myriad Genetics*, a case evaluating the legality of obtaining patents on isolated human genes, Lourie's scientific thinking got in the way. When the case came to the Court of Appeals, Lourie wrote that cutting chemical bonds to extract a DNA sequence from its native place in the chromosome was enough to make the DNA sequence chemically novel. As such, he wrote, it became an artificial material "eligible" for patenting. The Supreme Court reversed him, pointing out that the issue of patenting genes is not one of chemistry but one of law. Chemical novelty, implied the Supremes, is not the same as patent eligibility.

The *Myriad Genetics* case taught me that no matter how successfully we scientists turn ourselves into lawyers, we may carry hidden baggage that has to be watched carefully, if not shed altogether. Science and law are a tricky mix and even the best, like Judge Lourie, may sometimes miss a step.

# Part II

# Issues of First Impression

The age of commercial biology has had to deal with legal issues more complicated than those that the wandering artisans who escaped Constantinople brought to the Venice Council 500 years ago. We are no longer confused about how to patent the building of canals or the weaving of silk. Our modern courts and patent offices have had to move into new territory and resolve cases with issues never seen before.

What are the legal differences between owning intellectual and owning tangible biological property? Who owns what is in your body or in your mind in the first place? Can you patent living things? Can you patent genes? How about purified natural materials? How do biotech inventors properly describe living beings, so they meet the Venetian requirement of full disclosure in exchange for exclusivity?

These are "issues of first impression," as jurists like to say. Let's see what happened when classical IP law was forced to adapt to the inventions of the new science.

# 6

# Who Owns Tangibles Taken from Your Body?

A few years ago, I was being rolled in for surgery to remove a suspicious nodule on my thyroid gland. Sensing the inexorable advance of the anesthesia and just for kicks, I briefly thought of raising an issue with my surgeon: Who would own my thyroid once it had been removed? What would the hospital do with it?

I was by then an experienced IP lawyer, well versed in questions of ownership of body tissues and of ideas, the tangible and intangible products of our bodies. And I was also aware that I had just signed a modern informed consent form, giving away ownership and commercial rights in my thyroid gland. The answers to my thoughts were therefore academic. They were the muddled musings of a patient about to go under the scalpel. The next thing I remember was waking up and a smiling surgeon telling me that all was well. I knew and didn't care by then that the thyroid gland was no longer mine.

These pre- and postoperation ponderings, however, did not always have as certain an answer as they had for me that day. For a long time, it was not that clear who owned property taken from our bodies.

## Tangible or Intangible Property

We tend to think of property as something that we can touch with our hands: a car, a pen, a piece of land. We call such property tangible. But something intangible, like an idea? How is that property? And, how can we say to whom it belongs, especially once it has been turned from a thought, secretly hidden in your brain's neurons, to a public utterance? Once it goes from your mind to the world, doesn't it belong to all?

Well, to quote Sportin' Life in *Porgy and Bess*, "It ain't necessarily so." The image of something inside one's body, and who owns it when it passes to the outside, is a good metaphor for both tangible and intangible properties of human origin. The initial condition is the same for both: if it is inside your body, it belongs to you. No one should have any serious argument with that concept. Well, maybe forcibly extracting an organ or a thought, a dreadful propensity of humanity's worst actors, are exceptions. A state doing such a horrendous thing in clear violation of the unwilling donor's autonomy believes it has the right to own the thought or organ even while still inside.

All civilized countries today recognize that whether it's a kidney or a secret formula in a person's head, both belong to that person. That is, so long as they remain inside. The question is what happens once an organ or anything else tangible, like a blood cell, or intangible, like an idea, is *voluntarily* brought into the world? Who owns it then?

To better understand the concept of intellectual property, such as an incorporeal asset that originates in our mind, it is useful to contrast it with the ownership of other things that come out of our body, such as blood cells, proteins, or DNA. We will start by discussing ownership of tangible human property. This will take us to the question of whether providing tangible materials to someone who then conceives an invention using them makes the provider of tangible property a coinventor of the intellectual property. In the next chapter, we will move on to the ownership of intangible things, like ideas and creations.

## Spleens and Cell Lines

A 1990 decision from the Supreme Court of California, *Moore v. U. California*,[1] gives us some answers to the question of ownership of human body parts. The facts and analysis of the *Moore* case are startling, so fasten your seatbelts.

In 1976 John Moore, a Seattle resident, was diagnosed with a rare form of blood cancer known as hairy-cell leukemia. This cancer gets its quaint name from the hairy look of a patient's malignant cells when under the microscope. Moore received treatment at the UCLA Medical Center under the supervision of Dr. David Golde. Dr. Golde recommended that Moore's spleen be removed, and, with Moore's consent, the operation was carried out in October of 1976. For almost seven years afterward, Dr. Golde directed that Moore return to UCLA for follow-up visits. In each of those visits, Golde obtained samples of blood, serum, skin, and bone marrow from his patient. The doctor told Moore that these procedures could only be done in Los Angeles under Golde's supervision, not in Seattle. What Moore did not know, because he wasn't told by anyone, was that Dr. Golde arranged with Shirley Quan, a scientist at the University of California, to share portions of Moore's spleen. She would then do research on its contents.

Around 1979 Golde and Quan established an immortalized cell line from the lymphocytes in Moore's spleen, one that produced on demand valuable amounts of immune substances known as lymphokines. The cell line was christened by the University scientists as the "Mo" line in reference to its origin in its patient Moore. Lymphokines are natural agents against cancer, and useful as biological drugs. An immortal cell line producing lymphokines on demand in a pharmaceutical lab is a valuable piece of tangible property. Two years later, the university applied for a patent on the cell line, listing Golde and Quan as inventors; the patent issued in 1984.[2] It was estimated that by 1990 the yearly market for the valuable lymphokines produced by the Mo line would be about $3 billion. Because of the university's patent policy, the university, Golde, and Quan would share in any royalties or profits arising out of commercialization. Sure enough, like bees attracted by the sweet nectar of a flower, biotech companies and their deep-pocketed partners in the big pharma world came buzzing. The university negotiated exclusive agreements with Genetics Institute and Sandoz Pharmaceuticals for commercial development of the Mo line.

Moore did not know anything about this until seven years after his surgery. What gave it away was that in 1983, during one of his follow-up visits to Los Angeles, he became suspicious of a new informed consent form he was asked to sign. The new form said, "I (do/do not) voluntarily grant to the University of California all rights I, or my heirs, may have in any cell line or any other potential product which might be developed from the blood and/or bone marrow obtained from me."

This form was different from the original consent form Moore had signed before his spleen operation in 1976. The first one had simply said that UCLA Medical could "dispose of any severed tissue or member by cremation." The new one mentioned things like cell lines and other "potential product[s]." It sounded more like a transfer of property to the university than an informed consent form. Suspicious, Moore gave the new form to an attorney, who discovered the patent and the commercial arrangements.

Moore started litigation. He sued Dr. Golde, Quan, the university, Genetics Institute, and Sandoz. He asked the courts for money damages based on several theories, only two of which survived the long road to the Supreme Court of California. The first theory was based on lack of informed consent, and the second one was based on what is known in law as "conversion." Let's discuss conversion first, since it is as clever as it is entertaining.

Conversion is a legal wrong that has been made to a person by taking away their tangible property and asserting ownership over it that is inconsistent with the real owner's title. It is the civil law equivalent to the crime of theft. In a criminal complaint of theft, the state's prosecutor brings an action against the thief for some sort of penalty. In a claim of conversion, it is the owner who sues the converter for damages. Conversion is an ancient cause of action, going back to England's early common law. A usual conversion in those days was for someone to cut down trees from someone else's land without permission, haul them away, and sell them as lumber. Another common conversion was taking cloth from its owner, also without permission, and making and selling garments. The critical aspect of the law of conversion is that even after the trees or the cloth have been removed from the owner's possession, they still belong to the owner. So, transforming them into lumber or garments is legally irrelevant.

It is unfortunate that the legal name for this offense is conversion. The concept has nothing to do with the *transformation* of trees into lumber or cloth into clothing. The word comes because the taker of the goods has "converted them" to her own use. The legal remedy for conversion was that the cut lumber and the finished garments still belonged to the original owner. If the property had already been sold, the original owner was entitled to their value in money.

In analogy to these ancient causes of action, Moore alleged that Dr. Golde and his collaborators took his blood and serum and "converted" them without his permission. He argued that he continued to own his cells following their removal from his body and that he never consented to their use in lucrative medical research. Just as for the hauled-away lumber or garments, Moore argued that the results of the wrong he had suffered were the Mo cell line and the patent. He said that they were also his, although he did not ask for them to be returned. After all, what would he do with a cold petri dish of his cells? What he wanted was to share in the money made from their use.

This complaint was so unusual that the Supreme Court of California found no case or law directly on point. It was, as judges say, a "case of first impression." To reach a decision, Judge Edward Panelli, writing for a majority of four members of the court, analyzed and tried to apply a wild array of precedents from remote areas of the law. The court looked at legislation dealing with who owned body parts that have been removed from a body. The judges looked to laws dealing specifically with disposition of human tissues, transplantable organs, blood sales, or aborted fetuses; or the use of extracted pituitary glands, corneal tissue, and even cadavers. They looked to California statutes meant to protect public health and safety, dealing with excised anatomical materials, tissues, bodily remains, or infectious waste. Nothing applied directly.

In a bit of a Hail Mary, Moore's lawyers cited a 1976 case from the state of Maryland, which dealt with the—one can only hope, very careful—seizure of a criminal defendant's poop from a hospital bedpan by police searching for narcotics. In that case, the Maryland court had held that the hospital patient had abandoned his poop, and that it was no longer his property. Its seizure by the cops didn't violate the US Constitution's Fourth Amendment prohibition against "unreasonable search and seizure." If strictly applied to the Moore situation, the Maryland case would have gone against Moore, since, just as the unfortunate crook had abandoned his poop, Moore could have been seen to have abandoned his spleen. In making its decision on the hospitalized delinquent, however, the Maryland court had floridly commented in passing that it was not unheard of for a person to assert ownership over such things as "fluid waste, secretions, hair, fingernails, toenails, blood, and organs or other parts of the body . . ." The Supremes of California called this passing comment a "slender reed" on which to decide the Moore case. They dismissed application of the Maryland court's comment to Moore as that case was related to a criminal procedure on suppression of evidence. Moore's, in contrast, was a civil dispute over who was entitled to the economic benefit of his removed spleen. Different area of law, and not applicable, said the California court.

The court nevertheless concluded that all these laws and precedents drastically limited Moore's control over his excised cells. It decided that Moore stopped owning his cells once they were no longer inside his body. The cut trees and the cloth of old were still the property of the landowner and the cloth maker, before they were improperly taken away, but not so with Moore's cells. They were no longer his once the surgeons took possession. Ultimately, the court disagreed with Moore that he had a cause of action for conversion. Moore never had legal title to his tangible spleen cells once they were outside his body. Consequently, UCLA Medical, Golde, or Quan did not "convert" anything to their own use.

Despite its conclusion that Moore had no cause of action based on conversion, the court was clearly offended by what his doctors did to him. They misled him, said the court. They urged him to return to LA for years of additional procedures, and they hid from him that they had a financial interest in research with his cells and in the commercial products that might come of it. A doctor has a fiduciary obligation to his patient, added the court. This includes telling a patient that the doctor might benefit commercially from a procedure he is recommending. Moore was entitled to know, before he gave his consent to the spleen removal, that Dr. Golde had personal interests, whether research or economic. These interests, which were unrelated to Moore's health, might have affected the doctor's medical judgment. Since the consent Moore gave was not fully informed, the Supreme Court held that it violated the doctor's fiduciary obligations.

The case was sent back to the lower court to reevaluate the evidence and, as courts of appeal like to say in their elaborate judicial language, "to render a new decision not inconsistent with our opinion." However, the case settled without going through another trial and without a new decision from the lower court. This is not an unusual result in our legal world. Sometimes we are left without knowing exactly what happened to the parties involved. Yet it is the analyses of appellate courts, such as the Supreme Court of California, that establish precedent and serve as guideposts for similar disputes in the future.

Moore's cancer went into remission due to the timely and well-advised removal of his spleen by Dr. Golde in 1976. He died in 2001, twenty-five years after his operation. Dr. Golde saved his life with the timely surgical intervention. Golde was a good physician but, unfortunately, a confused pioneer in the then new frontier between biology and commerce.

The *Moore* case has been followed throughout the land, not only in California. It stands for the simple proposition that if you are to be operated on to remove, say, your appendix, once it's outside your body it is no longer yours but belongs to the hospital. That is the law, unless, of course, you sign a contract with the hospital ahead of time that even when outside your body, the severed appendix will still be your property. That is unlikely to happen. After *Moore*, most consent forms signed by patients before operations contain waivers of ownership and of any rights to commercial benefits. Such was the consent form I had signed before being rolled into my thyroid surgery.

## Turning Tangibles into Intangibles

Let us compare the ownership of tangible body parts with the ownership of the patents on intangible ideas that result from the contribution of the tangible parts. The Supreme Court of California considered and then dismissed Moore's claims that he had ownership rights over the patent, that is, over the intangible property derived from his cells. The judges explained that the patent law rewards inventive efforts, not the provision of naturally occurring raw materials. The Mo cell line and its patent were the product of invention and, in principle, belonged to the inventors, not to Moore.

The court recognized that for inventorship of the patent, Moore needed to have made an *intellectual* contribution to the mental act of conception of the cell line. This he did not do, said the court. The conception that led to the patent on the Mo line was purely that of Dr. Golde and his collaborator Shirley Quan. All John Moore contributed, said the majority in the California case, were the raw materials, not the brainpower.

There was an eloquent dissenting opinion, written by Judge Stanley Mosk. He said that although a patient who donates critical cells does not fit squarely within the definition of an inventor, good public policy should lead to at least benefiting him financially. Moore's claim to share in the profits flowing from the patent would be justified, said the judge. Mosk's views of what would be good and fair, however, did not carry the day. He convinced only two more judges to go along with him. The law of inventorship was then, and remains to this day, based on intellectual contributions, not contributions in-kind. Yet nothing would have prevented UCLA Medical from willingly sharing a portion of its financial gains with a patient whose cells were so valuable.

A similar situation arose in the case of Henrietta Lacks and Johns Hopkins Hospital in Baltimore. In the early fifties, twenty years before John Moore's spleen was removed in California, Johns Hopkins biopsied Lacks to evaluate her for cervical cancer. She died without knowing that her cancer cells, which were obtained from the biopsy, were very useful and would soon be made into one of the most famous and valuable immortalized cell lines in the history of medicine. The line, named HeLa, is still around, and is used throughout the research world. Decades after Lacks's death, through the sleuthing of science writer Rebecca Skloot, Lacks's relatives discovered that their ancestor's cells had become immortalized. In 2011 Skloot wrote a powerful book about the story, *The Immortal Life of Henrietta Lacks,*[3] which was made into a movie.[4] The resulting outcry and publicity made a major impact on Hopkins and the scientific community.

Contrite, Johns Hopkins Medical School now has several events honoring Lacks's contribution of her cervical cancer cells. It hosts two annual Henrietta Lacks Symposia focused on community outreach and medical ethics; it established a Henrietta Lacks Scholarship, and it offers the Henrietta Lacks Memorial Award. Public shame goes a long way, especially when it casts a shadow on famous institutions.

In late 2021 the Estate of Henrietta Lacks sued Thermo Fisher Scientific (TFS).[5] TFS is a commercial lab that sells products derived from HeLa cells and its extracts.[6] Echoing the dissenting views of Judge Mosk in the *Moore* case, the Lacks descendants demanded that Thermo Fisher "disgorge" all profits made from commercialization of their ancestor's cells. The case settled in 2023, but the financial terms were not revealed.[7]

The *Moore* case dealt with a spleen in which the ownership changed when it went from the inside to the outside of Moore's body. Does the same happen with incorporeal things, such as an invention, a piece of music, or the mental image of a painting or a clever rhyme? These are the types of products of the mind that are at the heart of intellectual property. Does intellectual property change ownership when it goes from the inside to the outside, as is the case with body parts?

# 7

# Who Owns the Intangibles?

Sending intellectual property into the world is more complicated than removing an appendix, a gland, or a spleen. If you keep your thoughts to yourself, you run no risks. They will be yours forever. But few creative types are able to keep their thoughts in for too long. Most have the urge to share their creations as soon as possible, perhaps to get praise and support. Or they think that their creations will make them rich and famous. Either way, if they describe their idea to the public, orate a poem, or play their music to an audience, they risk losing ownership unless they take precautions. Ideas are so ephemeral that no one listener will own them after they come out of their creators' heads. They will belong to the world.

In 1976 Richard Dawkins, the English biologist, wrote *The Selfish Gene*, in which he advocated that all of us living beings, from bacteria to humans, are nothing more than exquisitely evolved carriers designed to make sure that our genes pass on to the next generation.[1] In his book, Dawkins proposed the concept of a "meme." A meme, which is meant to rhyme with "gene," describes a unit of thought that resides in the brain. In analogy to what genes do when inside, memes reproduce, mutate, and evolve but do so once they are outside. It's a neat little theory.

We could take the analogy a step further. Like expunging a human gene from the genome changes its ownership from the human whose body is the gene's source to the expunger, exporting a disembodied meme changes its ownership from the creator whose mind is the meme's source to all of us. Unless the creator does something proactive to make sure that her memes remain hers after the relocation, they will drift into the wind as so many spores. There, they will be copied and exploited by others.

The elaborate body of laws dealing with intellectual property tries to make sure that inventors, artists, writers, and other such types can own the creative memes that come out of their minds.

## Babies and Inventions

The IP world is usually divided in three parts: copyrights, patents, and trademarks. For the sake of this tale, we'll deal only with the first two. If the creations are literary, musical, or choreographic, copyright law applies. Copyright law establishes that once the creation is fixed in a tangible medium—such as a piece of paper, recording tape, or hard disk—it belongs to the creator; there's not much more to it than that. It helps to give notice to the world of your ownership by placing a phrase somewhere noticeable, such as at the beginning of a composition or book: "© Ernest Hemingway, 1952." But even that is not necessary. A creator of music or a poet is well advised to take the precaution of making his intangible creation tangible as soon as possible, before singing it out, as loud as he wishes, to the rest of us. He will still own it. The big exception to this rule is when the creator has been hired to create things, like write an ad or compose a jingle. In such cases, his creation is a "work for hire," and the copyright belongs to the one who pays the piper.

If the creations are technological, such as a machine or, more relevant to this story, a microorganism containing a man-made recombinant DNA molecule (rDNA), patent law comes in. The one who conceives of such a creation is known as the inventor, and the idea belongs to her. I have met many an inventor who compared bringing to the world an incorporeal creation of her mind with delivering a baby. There is credence to the simile between babies and intellectual creations, in that the very word in law for the inception of a technological idea is "conception."

Once the idea is delivered to the world, however, it may be lost to the creator unless, before opening her mouth or publishing a paper, she files a patent application. If she spilled the idea ahead of filing, it's too late. This is not as easy as with copyright law, where protection is available by simply writing down the phrase, rhyme, or melody on a piece of paper.

In patent law, there must be, before any public disclosure, an actual filing of an application in one of the many patent offices of the world. Once the application is filed, the inventor can freely talk to the world about her delivery. In the United States we give the inventor a one-year grace period before filing a patent application.[2] Most other countries of the world don't have grace periods.

The next question is who owns the ultimate patent once an application has been safely filed? The answer is it depends. Was the inventor self-employed, like a lone garage tinkerer, or was he employed by a company at the time of his conception? That may make the difference. If he conceived of it while employed by no one, he owns it. If, when he conceived of it, he was employed, he had likely already signed away the ownership of his IP to his employer. The fine print in most agreements of a company with employees who are hired to invent contains a clause transferring ownership of all future ideas to their employer. The courts have held that just as Moore's cells belonged to him until his surgeon removed them and they then belonged to UCLA Medical, the transfer of ownership of an invention made by an employee occurs the moment the idea goes from the employee's mind to the outside world. Such a moment is oftentimes way before the filing of a patent application on the invention. It doesn't matter: ownership of the invention is established a long time before filing for patents.

While hidden in the employee's mind, the invention does not yet belong to the employer.[3] The employee can quit, move somewhere else, and only then write of his creation. That becomes the moment when the conception's existence can be proven, and because it's after the departure, it no longer belongs to the company left behind. The whole issue turns on whether there is sufficient evidence to corroborate the, obviously self-serving, story by an employer that the employee conceived the invention before leaving, and that it belongs to the company, not the inventor. Usually it's a "he said, she said" dispute. Without hard evidence, such as a written record, the employer loses. If, on the other hand, the employee's conception had been written down before leaving, or a credible witness comes forward to corroborate the employer's side, the conception belongs to the employer.

Here is a story that illustrates a dispute about the ownership of intellectual property once it came out of a scientist's mind.

## The Sewers of Pasadena

In the late 1980s I had an amusing case involving a patent on a purified protein and its use in a diagnostic test for HIV-AIDS. I was defending my corporate client against an accusation that its use of the protein infringed the patent of the bad guys. I should say that we lawyers usually refer to our adversaries as the "bad guys," in contrast to our clients, who are invariably the "good guys." It's a judgement-free shorthand that makes discussions at our internal conferences much simpler.

The issue of who owned what came out of the inventor's mind was right at the center of the dispute.[4] The front cover of the bad guys' patent showed that the patent belonged to a small Michigan company. As I started investigating the origins of the patent, I discovered that about a year before the filing, the inventor had been a postdoc at Caltech, in Pasadena, California. Yet the patent owner was not shown to be Caltech, but the bad guys' company in Michigan.

I smelled a rat. I figured that the bad guy had walked away from Caltech, never telling the university that he had made this invention. Once in Michigan, he had filed a patent application and gave its ownership to his new startup company. I learned that Caltech had a clear policy that all inventions made using facilities of the institute belonged to the institute.

After learning all of this, I took the deposition of the inventor. I was loaded for bear. The deposition took place in the offices of his New York attorneys, high up in a skyscraper on Park Avenue and 38th Street. As I looked north, I could see Grand Central Station with the Pan Am building on top. Helicopters had not landed on the roof since 1977, when one of them crashed up there after landing. If all went well with my plans, the bad guys were about to crash too.

I started the deposition by asking the inventor if he had been a postdoc at Caltech during the years just prior to the filing of his patent application. He said, yes. Then I showed him a copy of Caltech's patent policy and asked him if he had ever seen it. He took his time but eventually said that, yes, he had seen it. I asked him to read the part that said that inventions made using Caltech facilities belonged to Caltech. He was silent for a while, but the slowly developing sweat on his brow gave him away. After I asked him to do so, he read the crucial portion of the policy into the record. I asked him if he had used any facilities of Caltech in purifying this patented protein. Hearing the question, the witness turned pale, and his lawyer started fidgeting with a pencil. He answered my question by saying, no, that he had done all his work in his garage in Pasadena.

It was a stunning answer. He was doomed, and I took my sweet time.

I referred to one of the examples in his patent, which described that the gene for the protein had been separated using what is known as a high-voltage gel electrophoresis machine. Now, in the 1970s and '80s, that was one expensive machine.

"Ok, so did you have a high-voltage gel electrophoresis machine in your garage?" I asked calmly. The sweat had now started appearing on his white shirt, and I knew that the video I was recording of him would show it. It is never a good idea for a witness to be videotaped wearing a white shirt. It makes him look pale, like he is hiding something. I always advise my clients to wear blue instead and a jacket on top.

His answer to my question about the machine in his garage was another stunner. "I don't remember," he said.

I was incredulous. "You don't remember if you had a High . . . Voltage . . . Gel . . . Electrophoresis . . . Machine in your garage?" I asked slowly, with as much sarcasm as I could.

"No," he said.

Moving right along, I asked him to look at the next example in his patent. This showed that one of his experiments had used molecules labeled with radioactive phosphorous, so-called P32. This stuff is poisonous and a highly controlled substance.

"Do you see the experiment?" I asked.

"Yes, I do," he answered in a voice that was barely above a whisper. I encouraged him to please speak loudly enough for the court reporter to hear him. Sensing danger, the lawyer asked for an interruption so his client could go to the bathroom. I said that I had a few more questions, and then we could all go to the bathroom together.

"Did you use P32 in your garage in Pasadena?" I asked.

"I don't remember."

I slowly pulled from one of my folders a document entitled "City of Pasadena Municipal Regulations." I showed it to him.

"Please turn to Section 3(A)(1)(c)(ii) [or something like that] and read along with me," I said. I read the part of the municipal regulation that prohibits any person who is a resident of the city of Pasadena from using any private or public drain to dispose of radioactive or other poisonous material into the city sewers. "Are you with me?" I probed.

He nodded. I reminded him that the court reporter needed to hear words and that nodding didn't do. He said, "Yes."

By now, the man was sweating like a sumo wrestler in a Turkish steam bath. His face was whiter than the Pasadena regulations.

In a final coup de grace, I asked, "Did you flush your radioactive P32 down the drain of your Pasadena garage?"

"I don't remember," he mumbled.

"Fine," I said. "Bathroom break."

The helicopter had just crashed.

During the break, I walked over to his lawyer and told him he should drop the case right then. It was a loser. If he and the client didn't drop it, his client would have to face the same questions in front of a judge, I added. Both of us knew how *that* would end.

The bad guys dropped the complaint a week later.

The conception of the purified protein belonged to the scientist all right, but the moment it was reduced to practice using the facilities of Caltech, it belonged to Caltech. The one thing that was clear is that it never belonged to the bad guys' company. Putting the patent in the name of his company derailed his case. He should have rightfully let Caltech own it, and then asked for a license. He was a good protein chemist but a lousy scammer.

## Posters and DNA

One last thing on how the ownership of intellectual property interacts with that of tangible property. The copyright law provides a neat illustration. If I buy a copyrighted poster, it is mine. And since it is mine, after I get tired of looking at it, I can resell it online or through a store. The copyright law does not prevent me from doing either. Copyright deals with the *intellectual* property that goes with the poster, not with the poster itself. But the intellectual property in the poster is not mine; it is the publisher's. Once the publisher gets its initial payment from me (say, through Amazon), the publisher's copyright is gone as far as getting more money from a resale. The publisher's rights are, as lawyers say, "exhausted by the first sale." Yet the publisher can stop me if I start making copies of the poster and selling them online. The copyright law prohibits me from going into competition with the publisher.

A similar thing happens with a test tube filled with patented recombinant DNA. The rDNA is legally different from a patent obtained on it. The rDNA is tangible property, and the patent on it is intellectual property. Just like a poster, the first sale of the test tube to me, the buyer, exhausts the patent rights of the patent owner. I now own the tangible rDNA and can resell the test tube to others. But the patent is someone else's, and the patent owner may prevent me from commercializing the rDNA inside. I can't make the rDNA from scratch and set up a small factory to sell it. I can do anything I want, short of going into competition with the patent owner.

So, if you bought them, the poster and the rDNA belong to you, and you are free to do anything you wish, to your heart's content. That is, except go into business competition with the copyright or patent holder. At the end of the copyright life, seventy years plus the life of the author, or the patent life, twenty years after the filing date, you are free to compete.

That is the generational bargain of the intellectual property laws. They delay public ownership for some time. The creator can own and profit from her creative memes for a while, but they will ultimately belong to the world.

# 8

# Quarreling Colleagues

An invention is a product of the mind. He who has a complete conception of an invention is the inventor. He may have collaborators who give him tangible materials or intangible concepts that help implement the original inventor's idea. If the original idea was complete, such collaborators are not co-inventors. But if the idea was not complete, then the collaborators may well be co-inventors. A central question with the invention of a biotechnological product is: Was the conception complete? Was something missing with the conception? And if so, did other people—in many instances longtime collaborators—contribute to make it whole?

Because so much follows from a mere mental act, determining if a conception is complete, by whom, and when, is among the most emotion-laden aspects of patent law. Most allegations of joint conception have two or more collaborators who have known each other for some time and who disagree as to whether both were needed—or not—to make a valuable conception complete. One will say that he alone conceived, the other will say no, that they both conceived it together. Perfectly collegiate friends can quickly turn into quarreling colleagues.

## Genealogical Trees

One such situation made its way into my practice in the 1990s. As I have described, I was for many years patent counsel to Mass General Hospital. I worked with one of their scientists to prepare, file, and eventually obtain a patent on the isolated gene for a fatal hereditary disease. Those were the days before the US Supreme Court, in the 2013 case of *Association for*

*Molecular Pathology* ["AMP"] *v. Myriad Genetics*,[1] told the biotech world that isolated genes could not be patented, as they were nothing more than information identical to that in their corner of the genome. We will get to that important decision in chapter 10. The patent I had obtained was still good when, one day, I received a call from an attorney in a big New York law firm. He told me that he represented someone who had been left out of the patent. He informed me that a detailed letter would follow.

The letter explained that his client had provided MGH with a genealogical tree of a large family, many of whose members had the disease. The same disease had shown itself in all its deadly cruelty in many of his client's relatives, he added. That led to intense interest by his client in other such families and to a collaboration with MGH's research team to find the culpable gene. His client had contributed valuable information, said the lawyer, and needed to be named as a co-inventor on the gene patent.

I took in all the information. I knew that the MGH scientist had traced the way the disease progressed through the genealogical tree. He had correlated each affected individual on the tree with unusual DNA patterns in their genes. After all that, he had been able to zero in on the precise region of the genome, from which he clipped out the portion of the nefarious DNA sequence. It was a scientific tour de force.

In the following weeks, I did a lot of legal research on the topic and talked to plenty of colleagues. I compared the contributions of the MGH scientist with those of the aspiring joint inventor. The contender's involvement was one step further removed from providing a tangible thing like a spleen, as John Moore had done. It was providing a genealogical tree that pinpointed family members affected with the symptoms of the disease. There were no further insights or intellectual contributions, just sporadic markings on the branches of the tree. In contrast, my client's scientist had done the work of thinking about and discovering the correlation between the markings and the small region of the genome that was malfunctioning. The patent was not on the tree but on the isolated gene. I concluded that delivering a genealogical tree was just like delivering spleen cells. It was not the type of *intellectual* contribution necessary to make the family researcher a joint inventor of the isolated gene. Neither delivery of the spleen nor the genealogical tree was enough to turn the deliverers into inventors.

The New York lawyer was not pleased with my response, of course. We went back and forth for several months, sparring about the law of inventorship and the *Moore* case. He threatened a lawsuit and implied public shaming. But my client backed me up—always a good thing. In the end, I think that I wore the lawyer down before he did it to me, and he went away.

But for John Moore's cells, or but for the availability of a large genealogical tree, the patents on the Mo line or on the isolated gene for the deadly disease may not have come to pass. Yet the inventors named in a patent are not determined by a but-for test for their in-kind contributions. The but-for question is asked but in an entirely different context: Would the invention have come to pass but for the *intellectual* contributions of this or that person? If the answer is yes, they are entitled to be named as joint inventors.

## Irate Professor

A similar situation involved a case I had in the early 2000s involving Professor Theodore Friedmann. When he retained me, Friedmann was an active faculty member at the University of California, San Diego School of Medicine. He is now emeritus but was then, and remains, one of the world-recognized pioneers in gene therapy, the science of replacing defective genes by functioning ones.

Professor Friedmann had been trying for years to convince the UC legal department to add him and one of his collaborators, Jiing-Kuan Yee, to a patent from which he felt they had unfairly been left out. A few years earlier, a different research group at the university, also working on gene therapy, had requested from my two clients—and they had provided—a small sample of material containing a gene encoding a natural "RB tumor suppression gene." The idea was that replacing in a patient a defective RB gene unable to suppress cancer by the normal gene able to do so would subdue the malignancy. The critical issue was that the provided sample was part of a genetic construction, a "vector" as it is called, that was uniquely useful for gene therapy. It was not just an isolated natural RB gene, or even blood containing the gene's DNA. The tangible sample contained a specifically engineered vector designed and created by Friedmann and Yee that could be used to introduce the normal version of the RB gene into cancer cells, thus replacing the defective gene.

UC filed for patents on a method of treating cancer using a gene replacement with the vector that Friedman and Yee had provided. But neither of these guys was named as an inventor. That seemed unfair to Friedmann and fishy to me. Friedmann was seeking legal advice and was angry enough to want to pay for it. Like the *Moore* case or the case with the genealogical tree, the Friedmann-Yee case also involved someone who had contributed critical material to another research group. The material was a test tube with a vector containing the normal RB gene, specifically designed for gene therapy. My legal question was: Did the Friedmann-Yee pair contribute only tangible material like Moore's spleen, or data like the drawing of a genealogical tree, both of which served as background materials useful for subsequent inventions like the Mo cell line or an isolated gene for a fatal disease? Or were the contributions of Friedmann-Yee *intellectual*, entitling them to be named as co-inventors? I thought that my client's contributions had a bit of both: they were tangible but with a lot of inventive content.

This time I was in the shoes of the New York lawyer representing the supplier of the genealogical tree. I called UC's legal department and complained about the fact that it had been giving my clients the runaround. The lawyer on the other side was sympathetic and helpful (not like me in the genealogical case). He agreed that the contributions of our clients were sufficient to make them joint inventors but that higher-ups at the university would not budge so easily. So, we put together a strategy by which both of us would tell the whole story to the Patent Office examiner in charge of the pending application. It was just the facts; we would not argue one way or the other. If the examiner agreed with me that Drs. Friedmann and Yee were co-inventors, the university would add them to the patent. If the examiner agreed with UC that they were not joint inventors, they would remain off and we would drop the case. We put together a joint statement in which UC did not disagree that the design of the gene vector in the test tube had come from the minds of my clients. We shook on it. Then we held our breath for a few months until . . . the examiner agreed with my clients! They were added, and their names appear on the face of the patent.[2]

Friedmann and Yee's contribution was very different from Moore's or the genealogical researcher. Moore was the provider of a tangible substance, and the genealogist was the provider of family information that led to the identification of the harmful gene. Both their contributions led to patents *but were not part* of the patented inventions. The patents were neither on Moore's spleen nor on methods of doing genealogy. The UC patent was on a method of gene therapy, and Friedmann and Yee invented a useful gene vector that was an integral *part of that method.* Their contribution was not just background material or genealogical knowledge. They were rightly joint inventors of the UC patent.

Dr. Friedmann was delighted with the legal result. He and his wife came to Washington and took Tim Shea, my law partner who had helped with the success, and me to dinner. That evening I met Friedmann for the first time. He was around seventy years old by then. He had blue eyes that, behind thin-rimmed glasses, sparkled with what, like the eyes of so many scientists into which I have peered over the years, was raw intelligence. He wore a subtle smile throughout the evening. The picture was that of a very bright man with a good sense of humor, who had left his anger behind. We all toasted our legal success.

In 2015, ten years after the issuance of his patent, Dr. Friedmann received the Japan Prize, not for anything to do with the patent but in recognition of a pioneering paper he had published in the journal *Science* in 1972, thirty years before we met. The paper, entitled "Gene Therapy for Human Genetic Disease?" raised for the first time the idea of replacing faulty genes in a human being.[3] The foundation that gave him the prize called him "the father of gene therapy."

I am not sure whether any royalties came our clients' way from adding the Friedmann-Yee pair to the UC patent. Gene therapy did not take off commercially as quickly as everyone, especially Dr. Friedmann, hoped. The best way to do gene therapy today is probably no longer to replace whole genes but to heal defective ones by the more efficient gene editing technique called CRISPR, which we discussed in chapter 4.

It is now clearly established that to be named as a co-inventor in a patent requires a *mental*, not a *physical*, contribution. Even if, as in the case of Professor Friedmann, his mental contributions came along with a tangible sample embodying his ideas, it was the mental part that made him a co-inventor, not the physical one.

## Snuggling Up

One of the hardest-fought battles over joint conception of a biotech invention was between two brilliant scientific teams from two different universities and two different countries. The case shows how detailed a task it is to parse contributions by collaborators and decide if a conception was complete or not. But having two academic institutions as a setting adds a twist to the inquiry. What if a researcher at one university partly conceived of an idea, but his deficient conception was only completed when he talked to a scientist at another university? Who owns the complete conception then? Both universities? The answer is yes. Such invention is jointly invented and, in consequence, jointly owned. Of course, if the invention is conceived by three or four joint inventors, a not-uncommon situation in modern biology, the analogy between the conception of a baby and the joint conception of an invention must part ways, lest the metaphor overstay its welcome. But, once a university employee, like the Caltech postdoc of the last chapter, uses the facilities of his university, the invention belongs to his university. If two scientists, each at a different university invent jointly, the final invention belongs to both universities.

The two quarreling teams in this case were involved in the most revolutionary cancer cures of modern times: immune therapy. The idea behind immune therapy is to use the body's own immune system to attack foreign cancer cells. The basic notion that an infection, and the accompanying immune response, can cure cancer has a long history. In the late nineteenth century, William Coley, an American surgeon working at NYU Hospital, inoculated a thousand patients having inoperable tumors with infectious *Streptococcus* bacteria or its products. Half of his patients showed complete tumor remission and survived for more than five years. About 20 percent of his patients had no clinical evidence of tumor twenty years after the treatment.[4] It took almost a century to understand what was going on.

In 2018 two scientists, James Allison from Texas and Tasuku Honjo from Kyoto, received the Nobel Prize in medicine for their elucidation of why the tumors in Coley's patients went into remission. It turns out that our bodies have a protective shield, manned by what are known as "immune checkpoints." When the checkpoints are functioning normally, they prevent the immune system from attacking our own cells, such as during pregnancy. Yet, when all is going well, the system is also unable to attack invading cells, such as cancer cells. But if we carefully poke a hole in the checkpoint armor, in a sort of *reversal* of the shield, we can release the force of the immune system against intruding cancer cells. The infectious bacteria of Coley were poking a hole in his patients' immune shield. And modern anticancer drugs, such as the monoclonal antibodies Keytruda® made by Merck, or Opdivo® made by Bristol Myers Squibb (BMS), have also succeeded in doing this. These so-called checkpoint inhibitors are used to reverse the armored immune shields of our bodies and send many previously intractable cancers into complete remission.

To give credit where it's due: the simile of poking a hole in the armored shield came from an oncology nurse at the Memorial Sloan Kettering Cancer Center, who, in 2018, explained it to me in such understandable terms. She was about to start an infusion of Keytruda into my wife's veins. Shortly after that, Keytruda sent my wife's lymphoma into long-term complete remission. Sloan Kettering is where Nobel Prize–winning Dr. James Allison did some of his fundamental work on immune check-

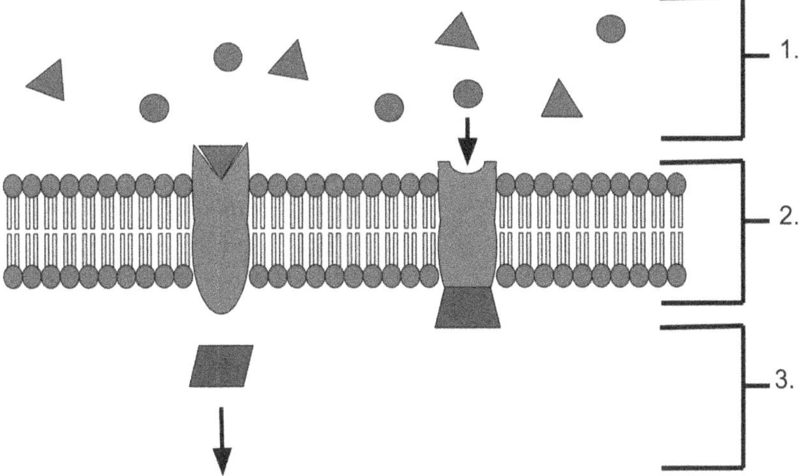


Figure 8.1. Two receptors embedded in a cell membrane. 1. Represents the outside of the cell with free-floating ligands. 2. Represents the cell membrane with a pair of receptors bridging the outside and inside of the cell. 3. Represents the inside of the cell with one signaling messenger bound to the receptor on the right and another signaling messenger released from the receptor on the left. Licensed under CC Attribution-Share Alike 4.0. Author Wyatt Pyzynski.

points. A 2020 movie, *Breakthrough*, tells of Dr. Allison's success in elucidating the role of checkpoint inhibition in cancer.[5]

Checkpoint inhibitors are some of the most celebrated drugs of our age. Of course, patents are never far behind in high-stakes discoveries. In 2002 Dr. Honjo and his collaborators in Kyoto filed an application asking to patent the use of monoclonal antibodies to unblock the body's armored checkpoints, unleash the immune system, and treat cancer. Their conception was based on the discovery of a receptor sitting on immune cells that, when bound to its ligand, suppressed the immune system. You may remember from chapter 4 the similarity I drew between a cell-bound receptor and a high school proctor sitting in front of a headmaster's office waiting for a troublemaker to walk by. Think of the proctor as the receptor anchored on the outside of an immune cell, and the ligand as the wandering troublemaker. The receptor recognizes the ligand, like a lock recognizes its key, and, after it does so, captures it and keeps it tightly snuggled. Once the receptor snuggles the ligand, the door opens for signaling. The signals let other parts of the immune system know that the lock has met its key, and the signaling suppresses the immune system.

Figure 8.1 shows a schematic of a cell membrane with the outside of the cell at the top and the inside of the cell at the bottom.

In the middle section of figure 8.1, which is numbered 2, is the cell membrane: this is the horizontal band of what look like pairs of soldiers opposing each other, standing guard across the figure, from left to right. At the top of the membrane is the outside of the cell (numbered section 1), with lots of ligands—some triangles, some circles—floating around. These are the troublemakers that would prevent a cancer cure. At the lower part of the membrane is the inside of the cell (numbered section 3), with two signaling "messengers" in darker shade, with different shapes. Embedded in the membrane (numbered section 2) are two different receptors. They are in a lighter shade of gray than the ligands at the top or the messengers at the bottom. The two receptors, like proctors sitting in front of the main office, have a connection both to the hallway outside and the principal's office inside. Of the triangles and circles floating outside the cell, only the triangles snuggle well into the receptor on the left with the triangular door, and only the circles snuggle well into the receptor on the right, with the semicircular door. Once a triangular ligand has snuggled into the receptor on the left, its snuggling releases a signaling messenger (the dark-shaded parallelogram in section 3) into the interior of the cell. The messenger, like a lone Pony Express rider carrying a pouch of instructions, sets things in motion. In this case the snuggling subdues the immune system and prevents things like autoimmune reactions.

If a physician injects her patient with a monoclonal antibody, the antibody may preclude the triangular ligand from snuggling into the receptor. The antibody, like a credenza jamming the door to the inside of the cell, "blocks" the receptor. The message to suppress the immune system does not get to the intended recipient inside, and the normal deterrence by the immune shield is disrupted. This disruption starts a cascade of events that may result in curing cancer—with the occasional side effect of an autoimmune reaction. Honjo and his Kyoto team had discovered one of such receptors, called PD-1, and on that basis filed for patent protection. They had the PD-1 receptor, say the one on the left of figure 8.1, and they knew how to block it with an antibody. They did not yet know for sure that there was a triangular ligand associated with the PD-1 receptor, but that did not keep them from filing for patents on a method of curing cancer.

So far I have simplified the explanation, but I'm afraid that now I need go a little deeper. As I noted, the receptor PD-1 (the "proctor") sits on the surface of immune cells. The triangular ligand, called PD-L1 (the "troublemaker"), is not floating around freely as I told you but oftentimes sits on another surface, that of cancer cells. When PD-1 and PD-L1 snuggle with each other, the cancer and the immune cells attach to each other like two boxers in a clinch. The immune cells, which start off by seeking out the cancer cells with the intent to destroy them, get quashed by the cancer cells *themselves*. Think about that: many cancer cells have evolved to suppress their natural enemy so that they are able to replicate at will. When a monoclonal antibody, like Keytruda or Opdivo, blocks the PD-1/PD-L1 interaction like a referee keeping the boxers from clinching, the suppression of the immune system is lifted. The immune cells can then attack the cancer cells.

The inventorship dispute arose out of the fact that the triangular ligand PD-L1, which is sitting on the cancer cells, was discovered by someone else, not by Dr. Honjo.

Dr. Honjo and his team obtained five patents in the United States. The patents were owned by Ono Pharmaceuticals and were exclusively licensed to BMS. Under the umbrella of these patents, BMS was the only pharma company that could sell monoclonal antibodies to the immune checkpoints. BMS started selling its own Opdivo for a variety of cancers, and then gave its competitor Merck a sublicense to sell Merck's Keytruda. Both drugs were estimated to control a worldwide market of close to $35 billion by 2023.[6]

Trouble for Dr. Honjo started in 2015. That year, the Dana-Farber Cancer Institute of Boston, where Dr. Gordon Freeman had his research lab, sued Dr. Honjo, Ono Pharmaceuticals, and BMS, demanding that Freeman be added to the patents as a co-inventor. Dr. Freeman was another world-class scientist who had been doing his own research on the inhibition of checkpoints to treat cancer. It turns out that Drs. Honjo and Freeman had been collaborating on this topic across the oceans for quite some time.[7] It was Freeman who discovered the ligand PD-L1 anchored on cancer cells.

The legal question in the lawsuit was whether Dr. Honjo had a complete conception of methods of treating cancer by using antibody checkpoint inhibitors or whether his conception was incomplete and was only completed after he received intellectual contributions from Dr. Freeman. If the lawsuit succeeded in proving that without Freeman's contribution Honjo's conception was incomplete, then Freeman would be added, and Dana-Farber would become a co-owner of the patents together with Ono Pharma. Since Merck was a license holder from Dana-Farber, Merck would no longer need the sublicense from BMS to sell Keytruda. A lot of money was at stake.

In 2020 the Court of Appeals for the Federal Circuit in Washington, DC, which decides patent cases, ruled in favor of Dana-Farber. The step-by-step analysis of whether Dr. Honjo's conception was complete or incomplete was done by Judge Alan Lourie, a scientifically trained jurist, whom I introduced in chapter 5 as an example of a scientist-lawyer. Lourie carefully dissected the scientific collaboration between Honjo and Freeman, the two brilliant scientists and their teams. The opinion by Lourie reads like a textbook of immunology. He deconstructed the invention and the intellectual contributions by both men, weighed what would be a complete conception, and concluded that Dr. Freeman was a joint inventor.[8]

Without getting as much into the weeds as the judge, I can say that his opinion explained that reversing the natural checkpoint shield was based on interfering with signals that immune and cancer cells send to each other. Such signals require the clinching and unclinching between a receptor anchored on the surface of an immune cell (the receptor in figure 8.1) and its ligand (the triangle), which in turn is anchored on the surface of a cancer cell. Judge Lourie admitted that Dr. Honjo had anticipated the role of the PD-1 receptor in anticancer signaling, but what was missing from a complete conception was the existence of the triangular ligand PD-L1. The existence of PD-L1 was the part that Dr. Freeman provided. Dr. Honjo had the lock, and Dr. Freeman had the key. Without Freeman's contribution of the triangular ligand that snuggles into Honjo's receptor, the conception of cancer treatment was not complete. A receptor sitting on an immune cell is just a receptor, said the judge. Without knowledge of its ligand sitting on the surface of a cancer cell the conception of curing cancer was incomplete. The result turned on the fine distinctions between a receptor and a signal. No ligand snuggling, no blocking of the signal; no blocking of the signal, no cure.

I am not sure that I agree with Judge Lourie. Since the biology world has known for decades about receptors snuggling with their ligands and generating signals, it was not hopelessly flawed for Honjo to foresee that there was a ligand out there; it was just a matter of finding it. You do not need the details of the key to prevent someone from opening a door. All you need to do is block the outside of the lock with some heavy furniture, which is the role played at the microscopic cellular level by a monoclonal antibody. If there's a lock, as Dr. Honjo discovered, it is likely that there must be a key that will open it. But short of getting your hands on the key and throwing it away, a hefty credenza that prevents access to the lock will do just fine. The credenza will not allow anyone who has the key to come over and open the door. That's what an antibody checkpoint inhibitor is: the blocking credenza. It prevents the ligand from snuggling into its receptor and opening the door. I might have ruled that Honjo's conception of treating cancer was complete even without proof of the triangular ligand. But then again, I was not the judge.

## He Said, She Said

Happily, I *was* the judge in an inventorship dispute in the 1990s, and it was a lot of fun. Two esteemed American institutions, a university, and a research agency hired me to be a private arbitrator and to help them resolve a legal disagreement. A pair of scientists, one from each, were in an intractable quarrel about their intellectual contributions to a biotech invention in the field of cancer therapy.

If you are starting to think that everything in the world of biotech patents revolves around cancer, you would not be entirely off. Curing cancer (or multiple different cancers, as we now know) continues to be the Holy Grail for medical research grants, university labs, biotech companies, big pharma, and venture capital. And all of them have their own patent attorneys.

Anyway, the two institutions wanted to file for patents but could not decide whether to name only one or both scientists. It was another classic "he said, she said" situation. One scientist, let's call him Dr. Me-Alone, held firm that he had had the whole idea, all by himself. The other one, called Dr. Jointly, countered that she had contributed a "Major Concept" to the idea, and insisted that without this, the invention was incomplete. Even though I charged by the hour, the cost for a private arbitration was many dollars less than if the institutions had gone to litigation. Plus, for much less money, they got themselves a scientifically trained arbitrator who understood the science as well as the laws of *intellectual* conception.

A situation where two people disagree—especially one where the question is: "What came out of whose mind?"—screams for proof. I asked both scientists to give us summaries of their contributions, accompanied by as much hard evidence as they could muster to support their side of the story. I have seen it all in my career: drawings on coffee-stained napkins, undated and indecipherable laboratory hieroglyphics, confused witnesses. But what Dr. Jointly produced took the award for clarity. Her packet of materials included an email she had received from Dr. Me-Alone a few months earlier. It said, in so many words, "Dear Dr. Jointly, I really enjoyed meeting you at the recent conference, and sharing our taxi ride to the airport. Thank you for the idea of the Major Concept you mentioned during the ride. I will be sure to use it when I do my next experiments." In addition, Dr. Jointly also produced dated handwritten notes that a member of the public took during a seminar that Dr. Jointly had given at her institution. The notes showed that way ahead of the taxi ride, Dr. Jointly had mentioned the Major Concept to several people.

I invited everyone to my offices. Both scientists were there, as well as their institutions' lawyers. I sat them on both sides of our long conference table, two on each side, facing each other, plastic water bottles for everyone. I was at the head, and Tim Shea, the partner who had done the Friedmann RB gene case with me and who had become a specialist in inventorship disputes, was next to me. I had forewarned the lawyers on both sides that I would rule in favor of Dr. Jointly. Neither of them was surprised and said that my task would be to convince Dr. Me-Alone of the reasons. Pride and ego are not unheard of in major scientific researchers. But if you add the possibility of gains from patents, the mix includes greed, and it can quickly become volatile.

I cut right to the chase. "After studying the record, your written submissions, and the evidence," I said, "I have concluded that the invention was jointly made by both of you. Both of you are co-inventors." I explained the need for corroborative evidence in these situations and how the email and the seminar notes had clinched it for me.

I looked at them. Dr. Jointly leaned back on her chair, not quite gloating but clearly relaxed. Dr. Me-Alone gave me a hard look, his elbows still on the conference table. The lawyers held their breath.

Suddenly, Dr. Me-Alone smiled gently, and said, "It was that ride to the airport. I had completely forgotten about that conversation, Mary. I am so sorry!" And with that, he leaned forward, reached across the table, and shook Jointly's hand.

You could hear the sigh of relief from the lawyers.

The careful dissection made by Judge Lourie of whether the complete conception of treating cancer with checkpoint inhibitors had to wait until Dr. Freeman confirmed the existence of signaling is only one of many such disputes that have bewildered jurists and lawyers for years. The reason is that the question of who conceived what and when is grounded in the elusive nature of thoughts and mental acts.

In 1972 Judge Clarence Newcomer of the Federal District Court in Pennsylvania expressed his frustration in having to decide if two inventors had collaborated sufficiently to complete each other's deficient conceptions. He said, in a sentence much repeated over the decades, that joint inventorship "is one of the muddiest concepts in the muddy metaphysics of the patent law."[9] If metaphysics is of interest to you, then litigating or arbitrating battles over sole or joint inventorship might be a good vocation.

But if you go there, be prepared. When long-standing collaborators quarrel over credit, the feelings get raw. Old friends become enemies, academic colleagues become less collegial, and mentors fight with their grad students. Your role will be more than that of a lawyer well versed in the details of intellectual conceptions. You will have to be scientist, jurist, and advocate, for sure. But your talents as an empathic psychologist will also be sorely tested.

# 9

# Patenting Living Things

An important issue of first impression, one that arose early in the development of the new biology, was whether it was possible to patent living beings. We lawyers now take it for granted that you can. But that was not always the case.

One of the first things that the Patent Office had to confront in the late 1970s was a flood of applications claiming inventions based on live beings. Until then, no one had seriously thought of patenting sprouting seeds, flying insects, or roaming cows; yet applications started arriving. While patenting living organisms is today a hallmark of biotechnology, the Patent Office, of all places, should have been excited and welcomed these imaginative developments. Yet it resisted.

## Brewing Beer

The story of patenting living things does not start with the advent of modern biotechnology. It starts a hundred years earlier with Louis Pasteur. The famous French chemist and microbiologist, one of the scientific luminaries of the nineteenth century, and the man who confirmed that microbes do not arise by spontaneous generation before he went on to invent pasteurization, was at heart a brewer. In the prime of his prolific life, he created the first vaccines ever for chicken cholera, anthrax, and rabies, any one of which would have been sufficient for a lifetime of glory. But even before then, Pasteur figured out a way to improve the fermentation of beers. And he filed for patents all over the world.

A few years before his work on vaccination, Pasteur had discovered that if the yeasts in beer mash, which convert sugars to alcohol during the process known as fermentation, were in contact with air, the fermentation slowed down. The reverse was true: if the beer was fermented inside a closed vessel in the absence of air, that is, "anaerobically," the fermentation was quicker. Through clever experiments, he concluded that if the yeasts became contaminated with airborne molds or bacteria, which Pasteur called the "organic germs of disease," they would not do their job properly.

So, sensing that America was a growing market for beer drinkers, he applied for and received several patents. Some of them were on the process of fermenting the brew under anaerobic conditions, and others on the live tools used for the process. On July 22, 1873, the Patent Office granted Pasteur, then the laboratory director at the École Normale Supérieure in Paris, US Patent 141,072. Its title is "Manufacture of Beer and Yeast."[1] In a move that can only be described as unwittingly prescient of what would come a century later, Pasteur's lawyer filed a patent claim for the yeast cells themselves. The claim said, "Yeast, free from organic germs of disease, as an article of manufacture." The Patent Office, without much ado, went along and granted him his request.

Shown below in figure 9.1 is the main drawing in Pasteur's patent. The image seems to be a hand-drawn composite of views that Pasteur was looking at under his microscope at separate times. It shows a side-by-side comparison of two types of yeast mixtures: on the right, the natural, germ-laden mixture, and on the left, the patented yeasts free from germs. The caption is in Pasteur's own words.

Frankly, it is not easy to see the difference between both sides, so I have added a vertical dividing line and an arrow and legend to help. The main distinction is the presence on the right of several slim, long, what we might call, for lack of a better word,"corpuscles." This somewhat archaic term is used to describe a free-floating cell, not one that is part of a tissue. Since Pasteur's is a nineteenth-century patent, I think the term fits nicely. The thin, stringy-looking corpuscles don't show up on the left

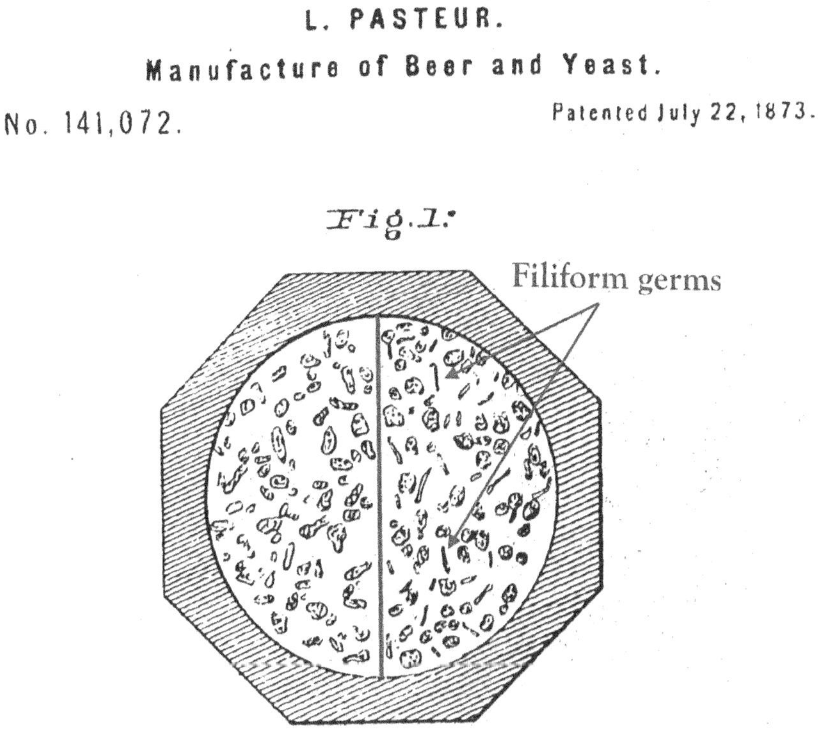

L. PASTEUR.

Manufacture of Beer and Yeast.

No. 141,072. Patented July 22, 1873.

Fig.1.

Filiform germs

Figure 9.1. A reproduction of fig. 1 from Louis Pasteur's US Patent 141,072 with the original verbatim caption: "The left half of the figure shows a pure alcoholic yeast and the other half an alcoholic yeast containing the diseased germs, which are filiform in appearance." Reproduced from the public domain.

and appear to be the culprits that slow down beer fermentation. Pasteur's words describing this figure confirm that the thin, long cells are the uninvited guests. He calls them "filiform," meaning "thread-like."

Pasteur was asking the Patent Office to give him a patent on the composition of yeasts as shown on the left, without the thread-like germs, and he got it. US Patent 141,072 is not to a process of brewing beer under anaerobic conditions. It is about the purified yeast composition itself, which includes a living organism. Mind you, the patent is not on any one individual yeast cell but on the overall composition, which is a mixture of live yeast cells and a few, if any, live germ cells. Pasteur's is probably the first US patent that includes a living being, and as such it is the first true biotech patent.

The phrase "free from organic germs of disease" in the main patent claim is there on purpose. It presupposes that naturally occurring yeast compositions are *not* free from germs. In his patent application, Pasteur did not ask for a patent on all yeasts; he could not have gotten it had he tried. Yeasts are a natural product, and even in the nineteenth century everyone agreed that you couldn't patent a natural product. No, Pasteur was asking for a patent on a subset of all yeasts: those that are not mixed with disease-causing germs. The careful language his lawyer chose limits the patent only to yeasts that have been freed of contaminating bugs. Pasteur and his clever lawyer hoped that they could convince the Patent Office that yeasts free from germs were somehow artificial, not natural. They even inserted into the patent claim the phrase "as an article of manufacture." These words are taken directly from the patent law and denote a novel and useful article made from raw materials. All of this was meant to convince the examiner that this subset of yeasts was most certainly *not* a natural product. And so, Pasteur got his patent.

The issuance of the patent was pioneering for at least two reasons. The first one was that it was to a living being. The second one was that the Patent Office considered that, living or not, yeasts "free from germs" were sufficiently different from yeasts in nature. These two legal issues are unrelated. You could try to patent live beings that have been entirely created in a lab and don't resemble anything that exists in nature. Take "Spider Goats," the genetically engineered goats invented by the Canadian company Nexia Biotechnologies in 2002, which produce spiderweb fibers in their milk.[2] Once extracted and purified from the milk, the fibers can be spun into steel-strength industrial cables. No such reprogrammed goats exist in nature. The legal issue here would be whether a live goat can be patented or not. The contemporary answer is it can.

In contrast, you could try to patent a non-live entity such as a purified protein, which exists in nature as a mixture with other materials. Among the most successful of contemporary medicines is the purified protein erythropoietin (EPO), a natural stimulator of red blood cells. EPO is manufactured by Amgen in genetically reprogrammed cells that carry the human EPO gene in their DNA. These are recombinant cells, and the EPO so produced is known as "recombinant EPO." The question with recombinant EPO is not whether it is alive or not. It most definitely is not; it is an inert molecule. The legal question here is whether purified recombinant EPO is sufficiently different from the EPO in nature, the one that appears in our kidneys to help generate red blood cells when we do not have enough of them in our blood. Natural EPO cannot be patented, but what about recombinant EPO? Maybe.

In the case of Pasteur's yeasts free from germs, both legal issues overlap. For one, yeasts are alive. And second, purified yeasts may or may not be sufficiently distinct from the natural ones to warrant a patent. Because of these questions, the issuance of Pasteur's patent proved controversial.

Almost forty years after it issued, in a 1937 article in *Science* magazine, P. J. Federico, a future examiner in chief of the Patent Office and one of the most respected scholars in the history of patent law, critiqued it.[3] He wrote that a patent on yeasts would "now [that is, 1937] probably be refused . . . since it may be doubted that the subject matter is capable of being patented." To a modern biotech patent attorney, Federico's critique is unclear because it seems to merge the two questions we have been pondering. He might have wondered whether yeasts, being alive, could or could not be patented. Alternatively, Federico might have thought that eliminating "germs of disease" was not enough to make such yeast compositions somehow artificial and not sufficiently removed from natural yeasts.

## Chomping Oil

The first of these two questions, "Can you patent living things?" was finally answered in the affirmative by the US Supreme Court in the landmark case of *Diamond v. Chakrabarty*, almost 100 years after the issuance of Pasteur's patent.[4] This decision held categorically that aliveness does not prevent patentability.

You might wish to ask: Why go to the Supreme Court? Doesn't the issuance of the 1873 patent to Pasteur on beer-brewing yeast establish that living things are patentable? The answer is no. If you cite a previously issued patent to the Patent Office as precedent and argue that your client is also entitled to a similar one, the answer is predictable. The Patent Office will say: "We may have made a mistake in issuing that one. And our previous errors are no precedent, sorry." Previous patents do not establish precedent. Only court decisions do.

Let's go back to being alive. In 1980, in a close five-to-four decision, the Supreme Court in *Chakrabarty* held that live genetically modified microbes capable of degrading a complex mixture of crude oils, as that which occurs in a spill, were eligible for a patent. The fact that these bugs were happily reproducing was no reason to deny a patent to Dr. Ananda Chakrabarty, the inventor, who was then a scientist at General Electric. Five members of the Court said that it did not matter that Chakrabarty's microbes were alive. The remaining four members of the Court thought the whole issue too controversial and would have had Congress deal with it. But they were in the minority, and the final decision was in favor of patents on living beings.

The Supreme Court acknowledged the contrary arguments made by Sidney Diamond, the commissioner of patents. Diamond, reinforcing the Patent Office's conservative stance, tried to convince the Court that living things should not be patented. The language the court used to address these concerns is so eloquent that it is worth quoting. First, evoking Hamlet, the Court described the trepidations of the Commissioner of Patents:

> [The commissioner] . . . points to grave risks that may be generated by research endeavors such as [Chakrabarty's]. The briefs present a gruesome parade of horribles. Scientists, among them Nobel laureates, are quoted suggesting that genetic research may pose a serious threat to the human race, or, at the very least, that the dangers are far too substantial to permit such research to proceed apace at this time. We are told that genetic research and related technological developments may spread pollution and disease, that it may result

> in a loss of genetic diversity, and that its practice may tend to depreciate the value of human life. These arguments are forcefully, even passionately, presented; they remind us that, at times, human ingenuity seems unable to control fully the forces it creates—that, with Hamlet, it is sometimes better "to bear those ills we have than fly to others that we know not of."[5]

Then, with a flair of its own, the Court answered the concerns of the commissioner, this time evoking Canute, the eleventh-century King of England, Denmark, and Norway:

> It is argued that this Court should weigh these potential hazards in considering whether [Chakrabarty's] invention is patentable subject matter. . . . We disagree. The grant or denial of patents on micro-organisms is not likely to put an end to genetic research or to its attendant risks. The large amount of research that has already occurred when no researcher had sure knowledge that patent protection would be available suggests that legislative or judicial fiat as to patentability will not deter the scientific mind from probing into the unknown any more than Canute could command the tides. Whether respondent's claims are patentable may determine whether research efforts are accelerated by the hope of reward or slowed by want of incentives, but that is all.[6]

The Supremes recognized that the DNA genie was out of the bottle and that, patents or no patents, it could not be put back any time soon. You cannot stop humans from investigating the unknown, said the Court, any more than Canute sitting on his throne by the sea could tell the waves not to wet his royal garments.

When they handed down *Chakrabarty*, the justices knew perfectly well that their ruling would facilitate the patenting of the many and clever inventions coming out of the then new biotechnology. The decision opened the legal floodgates. In short order, the Patent Office and the courts affirmed that all kinds of living subject matter could be patented if it was sufficiently modified by the human hand.[7] Dr. Chakrabarty patented his live microbes, and what followed were patents by others on human-made live seeds, plants, and animals. *Chakrabarty* posthumously reaffirmed the nineteenth-century prescience (or luck?) of Pasteur's lawyer that just because yeasts were alive did not mean they could not be patented.

So, we all agree now that being alive is not a bar to patents. If that were P. J. Federico's only worry in 1937 when he critiqued Pasteur's patent, his concern would have been settled by the Supreme Court in 1980. The alternative question of Federico, whether an invention like yeasts "free from germs" is no longer a natural product, was not fully clarified in the *Chakrabarty* case. The Court was not passing judgment on the validity of Pasteur's long-expired patent. What mattered to the justices in 1980 was that Dr. Chakrabarty's oil-chomping bugs were clearly not natural products. There was no evidence that microbes that could degrade mixtures of crude oil, any more than the Canadian Spider-Goats, existed in nature. There were microbes in nature that degraded oil, sure, but they did that one fraction at a time. The newly patented ones chomped at several fractions simultaneously, a big advantage over nature. They were artificial enough.

The Supremes in the *Chakrabarty* decision echoed another prescient move by Pasteur's lawyer. The lawyer called Pasteur's purified yeast an "article of manufacture." The Supreme Court also called Dr. Chakrabarty's engineered oil-eating bugs an "article of manufacture."[8] That made sense. Chakrabarty's wondrous oil-chomping bugs were microbes that were not naturally occurring. Chakrabarty had dramatically re-engineered nature.

The Supreme Court, however, did not deal clearly with the second question: How far removed from nature is removed enough for patent purposes? The Court agreed that the microbes of Dr. Chakrabarty were removed enough and worthy of a patent. Those were the facts of the case, and that's all the justices decided. Courts do not give legal advice; they resolve specific disputes. It's left to the rest of us to ask whether, just because reprogramming a wild microbe to do something never done before is eligible for patents, purifying wild bugs like Pasteur's yeasts to remove unwanted germs is also worthy. That is what I will call the "Pasteur question."

While the "Pasteur question" was not answered in 1980 in the *Chakrabarty* case, the Court approached it in 2013, in the case of *Myriad Genetics*.[9] In that case, which we discuss next, the Supreme Court held that even though isolating genes from the human genome involves the human hand required by *Chakrabarty*, isolated genes are not eligible for patents. And, while the *Myriad Genetics* decision did not frontally answer the "Pasteur question," it did give us some hints as to the answer.

# 10

# Patenting Genes

In 2013 the case of *AMP v. Myriad Genetics* presented the Supreme Court with profound scientific and legal questions concerning the nature of genes, whether isolated or not.[1] Are they chemicals, are they carriers of information, or are they both? These questions had already appeared to me forty years earlier, during my years as a graduate student at Harvard.[2]

In 1973, as a young doctoral candidate in organic chemistry, I took Professor James Watson's introductory course in genetics. I was there at the urging of my advisor Frank Westheimer, whose prescient view was that even for organic chemists, "the future belonged to biology." I wanted to be part of the future. Three times a week I left my research bench and walked over to the bio labs, its entrance flanked by the sculptures of Bessie and Victoria. These two massive rhino statues have stood guard since the 1930s, a time when biology was about big animals and plants, not their genomes and other microscopic entities.

Twelve years earlier Watson had won the Nobel Prize for his discovery—together with Francis Crick, Rosalind Franklin, and others—of the double helix structure of DNA. Now, he held forth on the biological roles of genes and their promoters, those segments of DNA that, like light switches, turn them on and off. While I was very impressed by my celebrity professor, the class bored me. I was an organic chemist. I wanted to think of the long DNA molecules as organic chemicals that could be manipulated in the lab.

Professor Watson, however, kept telling us that DNA molecules were information-storage elements, and he treated them as such. We learned that the genome was like a long winding road along which, as you walked forward and then back, you encountered shorter regions where you would find actual genes. The only difference between one gene region and another were the unique order of sequences formed from the alphabet of four chemical letters, A, T, C, and G. These sequences, while still lengthy, were much shorter than the complete genome, and formed, in countless permutations, the distinctive chemical sentences that encoded one protein or another. "This segment of the genomic DNA encodes for the protein insulin, and that segment over there encodes for the protein interferon," Watson might explain. I concluded that the closest these DNA molecules got to organic chemistry was that they were polymers, long chains of small chemical units. They reminded me of pearl necklaces. But the fact that they were such long chains didn't impress me much.

The tension between the views of DNA as information storage or as an organic chemical molecule would be played out dramatically one morning four decades later in the courtroom of the US Supreme Court. Dr. Watson was sitting in the front row of the public gallery, and I was sitting a few seats beyond him. The case before the Court that day was *Myriad Genetics*, and the issue was whether Myriad's patents on "isolated" *BRCA1* and *BRCA2* cancer-causing genes were drawn to subject matter that could be patented or whether the patents should have never been issued by the US Patent Office.

## Fuzzy Answer

Most manufacturing in modern biotechnology starts with natural sources. The natural world is the platform on which biotechnologists invent new and useful materials. Pasteur's yeast "free from germs" and Chakrabarty's oil-consuming microbes are but two of many biotech inventions based on nature. As we have seen, the 1980 decision of the Supreme Court in *Chakrabarty* held that aliveness would not prevent patent eligibility. *Chakrabarty* opened the doors to patenting microbes, plants, and animals. If they are artificial enough and legally distant from their natural counterparts, living beings can be patented.

Since patenting a natural material, like a preexisting bacterium or a yeast spore floating in the wind is a legal no-no, the degree of human intervention occupies the minds of biotech attorneys to this date. While we attorneys do understand that a natural material needs to be modified for patenting, the question is, How much?

The answer to that question remained fuzzy after the *Chakrabarty* decision. The Court was satisfied that the bacterium in *Chakrabarty* was clearly modified by the hand of a human and did not go into the issue any further. The justices did not evaluate how modified is modified enough. They called the bacterium a patentable "article of manufacture" and left it at that. But what about materials, live or not, isolated from nature? Can they be patented in isolated form?

In 2013 the Supreme Court handed down some wisdom about isolating things from nature, such as genes, and trying to patent them. In *Myriad Genetics*, the Court pondered whether clipping genes out of the human genome by breaking covalent bonds and trying to claim the genes in isolated form would or would not pass patent muster.[3]

Let's first touch upon some basics on the chemistry of covalent bonds, which are important to this story. Just a touch, I promise.

## Japanese Pearls

What are covalent bonds, and why are they important? Covalent bonds are the strong, hard-to-break chemical connections that hold together small molecules such as water, $H_2O$, and larger ones such as DNA. Each of the two hydrogen atoms in water is bonded to one central oxygen atom by just such covalent bonds: H-O-H. To break the covalent bonds in a liter of water into pure hydrogen and pure oxygen (like astronauts hope to do some day with water-containing Martian rocks), takes about 4.5 kilowatt-hours. That's a lot of energy. With that amount, you can run two to three loads of dishes in a kitchen dishwasher. Covalent bonds are tough to break, and they hold together $H_2O$ as well as the long chains of DNA that form the genomes of everything alive. Organic chemists spend their days obsessing on how to make or break covalent bonds in precise ways to make new molecules. When I was in grad school, I was very taken with covalent bonds.

In addition to evoking Watson's comparison of the human genome to a long and winding road, you might also think of the genome as a lengthy string of white Japanese pearls—the ones we met earlier, in chapter 2. Each pearl is attached to its two neighbors by covalent bonds. The inventors of the patents held by Myriad Genetics had spent laborious years searching the genome for the precise two spots where to cut the string to extract from it the much smaller segment that encoded the *BRCA1* gene, and nothing else. (They did that also for the *BRCA2* gene, but we'll go on with *BRCA1* for ease of discussion.) Once they extracted the DNA segment encoding the *BRCA1* gene, they could sequence it. And doctors, using the sequence, could figure out whether the *BRCA1* gene from a patient was identical to the one we all carry (and which would be unobtrusive to our life) or whether it had a deadly mutation (in which case it might increase our risk of breast cancer). Controlling the ability to commercially offer cancer diagnostics by owning a patent on the sequence of the *BRCA1* gene and its known mutations was big business for Myriad Genetics.

The Association for Molecular Pathology (AMP) had questioned the validity of the Myriad patents, challenging the notion that the mere fact of breaking covalent bonds and snipping out pieces of the genome encoding this or that protein were enough to distinguish such isolated gene segments from their identical natural counterparts, which were fully integrated into the genome. The morning in 2013, as I was sitting in the Supreme Court a few yards away from James Watson, I was there to listen to the arguments in *Myriad Genetics*. The justices, and the advocates for both sides, AMP and Myriad, were discussing patent law, although, listening to the debate, I easily imagined being back in the bio labs in Watson's class.

I had not seen Dr. Watson since I took his course. He had left Harvard to head the Cold Spring Harbor Lab in New York. I too left Harvard, a PhD in organic chemistry in my resume, and, after a short stint as a postdoc doing research on enzymes, went to law school and became a biotechnology patent attorney. It was as a young patent attorney that I was still thinking of isolated DNA segments as organic chemical molecules, never mind that Watson had told me years earlier that they were information units.

## Chemistry Reigns for a While

For decades, most of us biotech lawyers were convinced that breaking the strong covalent bonds that held a gene enmeshed in the chromosome turned the isolated fragment into a material that was artificial enough to make it eligible for patents. We thought that cutting the bonds made the isolated genes into "articles of manufacture." The US Patent Office thought so too, and, in the period from the late 1970s—when the modern biotech revolution started—to the early 2010s—when AMP challenged the practice for the first time—it issued thousands of patents on isolated genes. The courts also went along. As they routinely evaluated patents on isolated genes, they never considered that breaking covalent bonds and excising a gene fragment from the chromosome were not enough to make it eligible for patents, or that these were minor steps with no legal significance.

I confidently wrote or helped obtain patents on many isolated genes, such as those for Huntington's Disease (isolated by Dr. James Gusella),[4] and cystic fibrosis (isolated by Drs. Tsui Lap-Chee, Francis Collins, and others).[5] We all thought as organic chemists in those days. DNA molecules were chemicals that could be manipulated. The resulting fragments were chemically different from the identical sequences embedded in the long DNA chains of nature. Such chemical difference was enough to make the isolated materials artificial and eligible for patents.

Over the years, I also represented many patent challengers who attacked the isolated gene patents of others based on every possible ground, except that I never argued that they were natural materials and therefore not eligible. The main reason, of course, was that my clients also held patents on other isolated genes, and they did not fancy rocking the boat of eligibility, lest their own patents fall off.

Our biggest champion in the legal interpretation of isolated genes as chemicals was Judge Alan Lourie of the US Court of Appeals for the Federal Circuit. We met Judge Lourie in chapter 5, where I introduced him as a hybrid scientist-lawyer become judge. His written opinions were always anchored in rigorous science.

When the patent eligibility of Myriad's *BRCA1* isolated gene was first challenged by AMP in 2010, Judge Robert Sweet of the Southern District of New York, to my shock and that of most of my peers in the profession, ruled that AMP was correct and that such isolated genes were not eligible for patents.[6] To our collective astonishment, Judge Sweet said that as a matter of law, breaking covalent bonds doesn't make any difference. With a fair amount of scientific snobbishness, I remember thinking, "He does not understand chemistry; just wait until Judge Lourie sets him straight."

And Judge Lourie did not disappoint. In 2012, on appeal from Judge Sweet's holding, the Court of Appeals, in a decision written by Lourie, provided a master class in organic chemistry.[7] Lourie made the existence of covalent bonds the scientific centerpiece of his legal reasoning:

> Isolated DNA has been cleaved (i.e., had covalent bonds in its backbone chemically severed) . . . to consist of just a fraction of a naturally occurring DNA molecule. . . . *BRCA1* and *BRCA2* in their isolated states are different molecules from DNA that exists in the body; isolated DNA results from human intervention to cleave or synthesize a discrete portion of a native chromosomal DNA, imparting on that isolated DNA a distinctive chemical identity as compared to native DNA. . . . In other words, in nature, the claimed isolated DNAs are covalently bonded to such other materials. Thus, when cleaved, an isolated DNA molecule is not a purified form of a natural material, but a distinct chemical entity that is obtained by human intervention.[8]

He even included a drawing that showed, in telescopic magnification, the progression from a cell with a nucleus filled with chromosomes, all the way to a DNA double helix. Figure 10.1. is taken directly from the decision of the Court of Appeals.

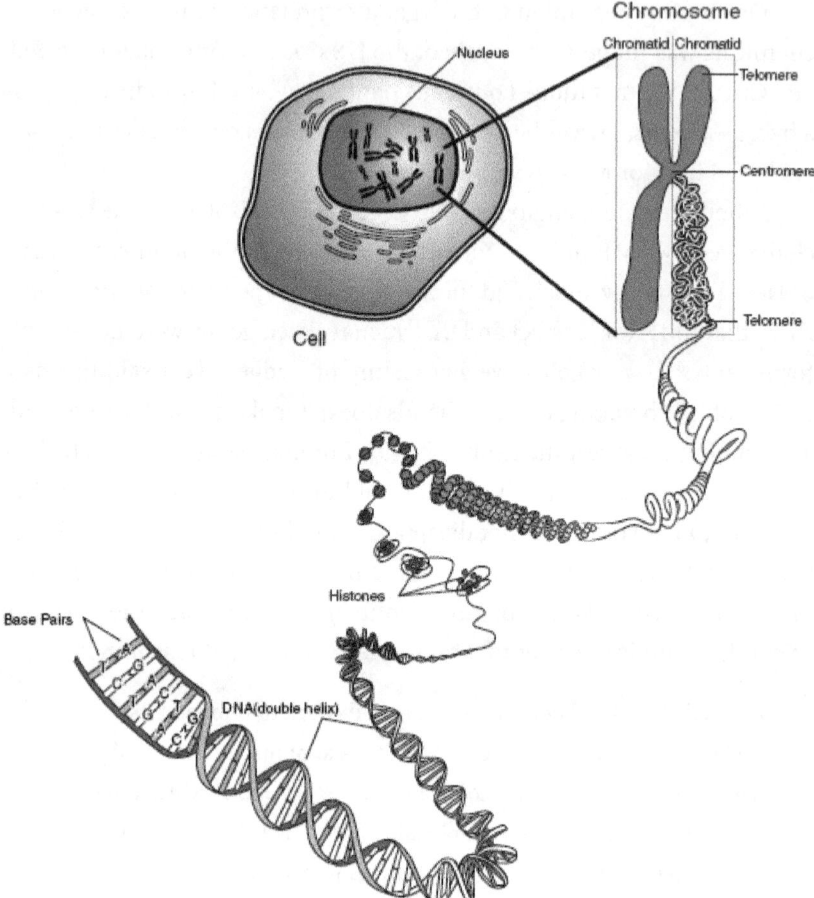


Figure 10.1. A cell showing its nucleus and, within it, several chromosomes that when uncoiled reveal the DNA double helix (not to scale). (Reproduced from fig. 3 of Association for Molecular Pathology et al. v. United States Patent and Trademark Office et al., 689 F.3d 1303, 1312 [Fed. Cir. 2012], a court decision in the public domain.)

I loved it. It was the perfect mix of science and law. DNA was a chemical, and covalently unmoored pieces of it were not products of nature. They were human-made "articles of manufacture," and patent eligible. The covalent bond stood supreme. Judge Lourie understood it and explained it for the ages.

Except the ages didn't last more than a year.

## Information Wins in the End

The case was immediately appealed to the US Supreme Court, and in 2013, the morning of the oral hearing, I found myself in the courtroom sitting half a bench away from Dr. Watson, my old genetics professor. I walked over and said hello, and he graciously said how nice it was to see me again. Coincidentally, I was also sitting next to a distinguished older gentleman whom I did not recognize. Since we were bound to wait another hour or so before the start of the proceedings, I decided to introduce myself and start a conversation.

"Good morning," he smiled back. "Very nice to meet you. I am Judge Robert Sweet, of the Federal District Court in New York."

"Oh my god!" I thought. "I am sitting next to the very judge who wrote the 2010 opinion in favor of AMP, holding that isolated genes are not eligible for patents! I better not tell him that I thought he knew no chemistry."

"I flew down from New York bright and early this morning," continued the judge. His excitement was contagious.

"Do you think that you'll be redeemed today?" I asked.

"Sure hope so. I knew that we would be here one day, and I wrote my opinion with that in mind. I didn't want to miss this day for anything in the world."

By then, almost fifty parties, in addition to the two principals in the case, had filed amicus curiae briefs, as friends of the court. Among them were many law professors, scientists, lawyers, public interest advocates, and professional associations backing one side or the other. Many were simply giving their opinion without backing anyone. The academicians expounded on the dual nature of DNA: it is *both* a long molecule *and* a storage information unit, they said, and this has generated confusion. One of them even pronounced rather glibly that the whole problem was that Judge Lourie of the Court of Appeals had confused science with law and that there was nothing legally special about breaking covalent bonds.[9]

I had also read Dr. Watson's amicus brief, which focused on the unique nature of DNA. As in 1973, he was fully on the side of information and minimized the view of DNA as a chemical entity:

> [T]he opinions by the appeals court miss the fundamentally unique nature of the human gene. Simply put, no other molecule can store the information necessary to create and propagate human life the way human DNA does. It is a chemical entity, but DNA's importance flows from its ability to encode and transmit the instructions for creating a human being. . . . [H]uman genes are much more than chemical compounds. . . . A human gene's patentability cannot depend simply on whether a covalent bond is broken during purification.[10]

By the morning of the hearing, sitting in the courtroom next to Judge Sweet, I had reluctantly come around to believe that the Supreme Court would reverse the Court of Appeals, rule against Myriad Genetics, and hold that isolated genes were not eligible for patents. A few months earlier I had given a lecture on the case at the University of Pennsylvania Law School. In preparing for the lecture, I read most of the fifty or so amici briefs. I was keenly aware that, in recent years, the Supreme Court had been on a streak, predictably reversing the Court of Appeals in patent cases. So, I told the Penn Law class that if I were on the Supreme Court, and along the lines of my hero Judge Lourie, I would affirm and hold that isolated genes are new chemical compounds and patent eligible. However, in hedging my bet, I also predicted that the Court would reverse. It was a win-win for me, and the students laughed along.

Half of me was right, of course. Justice Clarence Thomas, writing for the full Supreme Court, held that the mere breaking of covalent bonds is not enough for isolated genes to be legally "human-made." In essence, said the Court, plucking a line of 80,000 pearls (the length of the "isolated" *BRCA* genes) from the much longer continuous string of 3 billion pearls (the length of the human genome) doesn't distinguish the isolated line of pearls sufficiently from the same line with the identical DNA sequence as it has when in the original string. They encode the same information no matter where they are. The Court in *Myriad Genetics* also said that while genes isolated from the genome are not eligible for patents, certain other forms of genetic information, such as so-called cDNA (copy DNA) are eligible since they are manufactured in the lab and are sufficiently artificial.[11]

My opinion that the case should have gone the other way was irrelevant. This was now the law of the land. When I took Dr. Watson's course at Harvard, I had complained that genetics was insufficiently chemical, yet chemistry was the very point that the Supreme Court rejected. To them, as for Dr. Watson, DNA was storage for biological information. My doctoral advisor Frank Westheimer had also been right: the future, even the legal future, did belong to biology. New York's Judge Sweet had been redeemed. And sadly, Judge Lourie, my hero, had been wrong. The Supreme Court told him—told all of us—that we shouldn't confuse chemical novelty with patent eligibility; they are two different legal concepts. An isolated gene segment, while a novel chemical compound, one that is not present in nature as such, is not necessarily eligible for patents; it is just information. And you cannot patent information in a form that is identical to how it exists in nature. This is now a fundamental lesson in patent law.

Undeterred by its Supreme Court loss, Myriad Genetics went back to federal court a few months later, this time in Salt Lake City, and sued seven companies for patent infringement, based on almost twenty additional patents that had not been clearly vanquished by the 2013 decision of the Supreme Court. By then, however, the tide had turned. I was on the other side of its renewed battle, as the attorney for one of the seven companies. I joined in the attack of Myriad's patents as no longer eligible. Isolated gene fragments were nothing but disguised products of nature, I argued, echoing Dr. Watson. The Utah federal district court agreed, and the Court of Appeals affirmed his decision: Myriad's remaining patents would not likely survive our collective challenges.[12] Judge Lourie was no longer part of the three-judge panel that decided this case. There is no question, however, that now bound by the 2013 decision of the Supreme Court, had he been part of the panel he would have ruled against Myriad Genetics. In 2015 Myriad settled with all seven companies, and the "isolated gene" controversies ended at last.

The whole legal odyssey taught me a valuable lesson. Enamored with obtaining patents on isolated genes, I had overlooked that my background as a scientist, while helpful to inform and educate my views, should not necessarily be in control when I tried to predict the outcome of legal debates. As I always tell my law associates and students, the law is part logical reasoning and part public policy—and too much emphasis on one or the other risks leading you astray.

Oh, yes, I almost forgot. What about all those patents on isolated genes that I had written and so successfully defended in years past? Well, I am reminded of an apocryphal story about Abraham Lincoln as a practicing lawyer in the courts of Illinois in the nineteenth century. Lore has it that one morning Lincoln argued a contract case before the State Supreme Court and, before the court broke for lunch, won a decision for his client. In the afternoon, in arguing a different case, this time for the other side of a similar contract dispute, he took an opposite position than the one that had proven successful in the morning.

One of the judges was taken aback. "But Mr. Lincoln," goes the tale, "this morning you argued the exact opposite."

"Yes, your Honor," responded Honest Abe. "But this morning I was wrong."

I too had been wrong about patenting isolated genes. Yet with Judge Lourie by my side, I had at least been in good company.

## The Tricolore

One more thing. We learned in *Myriad Genetics* that an isolated gene claimed by sequence was not eligible for patents. But what if the gene after isolation is then recombined with an extraneous piece of DNA that has never been its next-door neighbor in the wild? Is the resulting "recombinant DNA" also not eligible? All of us in the patent community were holding our breath wondering whether the Supremes would say that recombinant DNA isolated in a test tube was or was not eligible. Much to everyone's relief, the justices didn't say that it was *not*.

Let's go back to the jewelry store we visited in chapter 2. You may remember that we there spliced together a string of Japanese white pearls with a string of Tahitian blues to form a two-colored necklace. Keep the image of an isolated human gene as a short string of white Japanese pearls in your mind. In nature, the string is linked on its left and right to other white pearls. If you were a microscopic observer sitting on the human genome, you would see a long stretch of nothing but white. Now imagine plucking out a white segment and, in a test tube, linking the left end to a string of blue pearls from Tahiti and the right end to a string of red ones from China. If you were now sitting inside the test tube at one end of that recombined string, and looking down along it, you would spot the Tricolore, that is, the blue, white, and red flag of France, as seen from the hoist side. Our tricolored pearl construction is analogous to a recombinant DNA molecule. In a test tube you might imagine a first segment of DNA from a bacterium, followed by a middle segment encoding a human gene, and, at the right end, a third DNA segment from, say, yeast. Is such isolated, human-made recombinant DNA artificial enough? What did the Supreme Court say about that in *Myriad Genetics*?

The Court said nothing. And that was good. Sometimes, reading a decision from the Supreme Court is like reading tea leaves: you need to understand what the justices *don't* say. The Court in *Myriad Genetics* walked a careful line and never said that an isolated DNA construction that includes a natural gene linked to a piece of extraneous DNA is *not* an "article of manufacture." This careful tiptoeing through the genetic tulips was subtle but clear. The biotech companies of the world had been isolating genes for almost forty years, and they learned from *Myriad Genetics* that they couldn't patent them anymore. They had been worried about such an outcome but not much.

What really kept them up at night was that the court might say in passing that isolated recombinant DNA *constructions* were not patent eligible either. That would legally muck up the industry's main patentable product, the engineered DNA that when inserted into bacterial hosts produces human proteins. The Court said nothing about that. The Court's golden silence told the scientists that they could continue creating something like a multicolored pearl necklace, and *that* they could patent. The day the decision in *Myriad Genetics* was handed down, you could hear a collective sigh of comfort from the genetic engineering industry. Everyone in the legal departments slept well that night.

Today we at least know a few more things than Pasteur and his lawyer did: Objects derived from nature, whether living or inert, can be patented if they are "artificial" enough. For that, they need to be modified in some way. The oil-eating bacteria of Dr. Chakrabarty were sufficiently modified, yet isolated genes were not; genes are just carriers of information, whether in the genome or in a test tube. But when genes are recombined with extraneous pieces of DNA, they become sufficiently modified to warrant a patent. So far so good.

Yet, even after *Chakrabarty* and *Myriad Genetics,* some issues remained unanswered. Frustratingly to patent lawyers, neither of these two Supreme Court decisions addressed the important question of whether pure natural materials other than genes are "artificial" enough to warrant a patent. Is the prohibition of *Myriad Genetics* on patenting isolated genes limited to genes, or does it extend to all purified natural products? What did *Myriad Genetics* do with the "Pasteur question," that is, whether a yeast "free of germs" is artificial enough to be eligible for patents? These questions are not just academic but have weighty legal and commercial significance.

# 11

# Patenting Pureness

Many inventions in biotechnology arise out of purifying natural materials. Think of finding a soil microbe that produces a valuable antibiotic, separating the bug from the soil, and extracting the antibiotic from the bug. Or think of an anticancer drug such as Taxol®, first revealed in the bark of a tree, at a time when USDA scientists were screening a thousand plants per year. Whether alive or inert, these materials are valuable as therapeutic agents or their sources. To make them successful, you must first find them and then purify them. And to raise investments for their production and distribution, hopefully you can patent them. And to patent them you need to answer the stark question: Is purity by itself a sufficient modification under the legal precedents to make a pure natural material patentable? That remains the core of our interest in Pasteur's question: whether purified yeasts or other such things that originate in nature would today be eligible for patent.

## Warfare

The cycle of searching, discovering, purifying, and patenting useful materials from nature is like Pasteur's nineteenth-century work on beer-brewing yeasts. Pasteur did not discover yeasts the way other scientists discover unknown microbes. Yeasts had been known since antiquity. But the French scientist discovered their impurities, and he received a patent on purified yeast. Removing impurities brought out the best in his beer-fermenting buddies.

*Chakrabarty* told us that a human-modified, oil-munching bacterium is well removed from a natural product. The genetic engineering had legally distanced it enough from nature that it could be patented. And *Myriad Genetics* told us that simply isolating genes is not enough of a step to provide legal distance. The biological difference between the materials in these two cases is that, in contrast to Chakrabarty's bugs, Myriad's genes were not modified genetically; they had just been isolated.

So, is that the line? Genetic alterations, yes; such inventions are patent eligible? Isolations, no; such inventions are not? The answer is not as clear as that.

The patent question of whether purifying a natural material is enough to make it eligible comes up often in the world of antibiotics. Historically, most antibiotics have been, and even today are, natural products derived from bacteria, fungi, plants, or animals. Researchers find them, purify them, and use them on humans. If they work, the inventors try to patent them. You might ask, "Why do bacteria have antibiotics? Is it to help us humans fight infections?" The answer is no, we have nothing to do with it. Bacteria didn't evolve to help *us*; we're just lucky bystanders. Let's explore this a bit.

In 1928 Alexander Fleming, working in St. Mary's Hospital in London, was doing experiments with a culture of bacteria when he noticed a green mold on one of his dishes. He also saw that where the mold was, the bacteria died. The next thing he knew, he had discovered penicillin, a natural antibiotic secreted from the aggressive mold. This discovery nicely illustrates the now accepted idea that we are living in a world of continual and unseen microscopic warfare. That's why bacteria and molds have antibiotics: to fight their microbial competitors. Fleming's molds have powerful weapons in their cells, which they use to destroy natural foes to efficiently compete for food and nutrients. Using these biological weapons for treating bacterial infections in humans is a sideshow.

You may remember another example of microbial warfare from chapter 2. There I told you the story of a bacterial protein called *Bt* that harms insects in which the unwanted bacteria make their home. The insects, in turn, are unwanted pests to several plants on which they feed. Plant Genetic Systems, a Belgian company, transferred the gene encoding *Bt* into the plants themselves and gave them protection against their insect predators. That would be analogous to inserting into humans the mold genes that produce penicillin and so protect us from bacterial infection. Such genetic engineering is too weird and dangerous to carry out, so it will remain a thought experiment.

Whether we genetically engineer humans or not, we should thank our bacterial colleagues that are fighting each other in the underworld. The enemies of our enemies are our friends. Consequently, we should keep looking for friendly antibiotics. We may hope to find additional ones by serendipity, like Fleming. In the alternative, we could methodically search every possible microbe, insect, plant, or animal we can get our hands on for useful hints that they harbor good antibiotics. Such a systematic search is what goes by the name of "bioprospecting."

Bioprospecting for new antibiotics able to combat ever more resistant bacterial strains is one of the central pharmacological pursuits of our time. The legal question is: Does finding and purifying a new antibiotic, such as Fleming did with penicillin, turn the pure antibiotic into an "article of manufacture"? Can you get a patent on it? The patent would be for the materials in unmodified form, with none of the genetic manipulations that Dr. Chakrabarty performed on his bugs. We would just purify them, that's all.

In analogy to Pasteur's patent on "yeast free from germs," maybe in our case the patent claim would say something like, "Antibiotic Z, free from natural impurities." Would that work? The antibiotic industry sure would like the answer to be yes. Without good patent protection, private companies hesitate to invest. Everyone is waiting for a big case from the courts that will, once and for all, answer the question, hopefully in the positive. While the *Myriad Genetics* case shed a glimmer of light on the issue, it did not answer it head on. For the time being that's all we have from the Supreme Court, so let's look at it more closely.

## Silence Is Golden

The Supremes in *Myriad Genetics* didn't say anything about purifying natural materials. Ironically, their silence on this issue was golden. At least they did *not* say that purifying natural materials does *not* lead to an article of manufacture. All that the Court said in its written opinion is that trying to patent an isolated gene by no more than its original DNA sequence is not legally removed from the same sequence in the natural genome. The sequence does the same in nature as in isolation, regardless of how pure it is. The DNA sequence carries information meant to build a protein, the same protein. Nothing has changed; no new function or property has been created by isolating DNA sequences.

Can we say the same thing about purifying a natural antibiotic or anticancer drug? Can we say that nothing has changed from its state in the wild to its state as a concentrated pharmaceutical? I don't think so. Certain discussions during the day of the oral hearing in *Myriad Genetics* show that the justices were aware of these issues yet said nothing about them when they wrote up their conclusions a month later. Because of this I believe that the Supreme Court would agree that *Myriad Genetics* does not prohibit patenting purified natural materials. You just need to know how to write your patent properly.

Those of us present at the Supreme Court during the *Myriad Genetics* hearing witnessed some interesting give and take between Justices Ruth Bader Ginsburg and Samuel Alito on one hand, and Christopher Hansen, the lawyer for the challenger AMP, on the other. Ginsburg started the questions in her inimitably gentle yet determined voice:

> Mr. Hansen, [your opponents say] that isolating or extracting natural products, that has long been considered patentable, and give—examples were aspirin and whooping cough vaccine. How is this [that is, an isolated gene] different from . . . natural products?[1]

Hansen did not give her a clear-cut answer. So, Alito came back to it a little later and asked, skeptically:

> Suppose there is a substance, a chemical, a molecule in the leaf—the leaves of a plant that grows in the Amazon, and it's discovered that this has tremendous medicinal purposes. Let's say it treats breast cancer. A new discovery, a new way—a way is found, previously unknown, to extract that. You make a drug out of that. Your answer

> is that cannot be patent—patented; it's not eligible for patenting, because the chemical composition of the—of the drug is the same as the chemical that exists in the leaves of the plant. . . . It's not just the case of taking the leaf off the tree and chewing it. Let's say if you do that, you'd have to eat a whole forest to get the—the value of this. But it's extracted and reduced to a concentrated form. That's not patent—that's not eligible?[2]

Hansen was better prepared for what I call Alito's "Amazon Forest" question. He answered quite assuredly:

> No, that may well be eligible, because you have now taken what was in nature and you've transformed it in two ways. First of all, you've made it substantially more concentrated than it was in nature; and second, you've given it a function. If it doesn't work in the diluted form but does work in a concentrated form, you've given it a new function. And the—by both changing its nature and by giving it a new function, you may well have a patent.[3]

These illuminating exchanges show that everyone in the courtroom that day was aware that equating the isolation of genes with the purification of natural products might lead to a result with which no one would be happy. Unless the Court carefully distinguished isolated genes from purified natural products, it was possible that neither would be eligible. Not even AMP's own lawyer was ready to go that far. In the end, Justice Clarence Thomas in his opinion concluded that isolated genes were not eligible but stayed short of wading into the tricky river of purified natural products. He wisely left crossing *that* bridge for another day.

Surely someone will discover a new anticancer drug by bioprospecting the world's forests and, after purifying it, will try to patent it. She will then face two hurdles. One is the bridge left uncrossed in *Myriad Genetics*, and the other is how to demonstrate that a purified antibiotic is, in the words that the Court uses to try and distinguish between natural and artificial, "markedly different" from the one found in nature. Until that day comes, let's reflect on forests a little more. They can teach us a lot.

## Yew Trees

The powerful drug paclitaxel, known as Taxol, which is useful for treating many forms of cancer, was first discovered by bioprospecting forests. In 1960 the National Cancer Institute (NCI) commissioned USDA botanists to collect samples from about a thousand plant species per year and screen them for anticancer activity.[4] In the mid-1960s, an active ingredient was found in the bark of a single Pacific yew tree in a forest in Washington State.

You might ask, What's an anticancer drug doing in a yew tree? Good question. The answer is it's doing nothing oncological. Its main function in the bark seems to be to protect the tree from invading fungi.[5] It is speculated that the famous endurance of ancient yew trees, which are known to live up to 2,000 years or more, is due to resident fungi defending against invading fungi by attacking them with paclitaxel. Its longevity is why the yew tree is a favorite of English cemeteries: a symbol of immortality. It turns out that fungi are fighting each other with chemicals that are useful to treat cancer in humans. Who would have guessed? It's that wondrous microbial warfare again.

Figure 11.1 shows the imposing Crowhurst Yew in St. George's parish churchyard, Crowhurst, Surrey.

In 2002 Queen Elizabeth II designated this tree as "One of Fifty Great British Trees." The Crowhurst yew is believed to be about 4,000 years old. That's the middle of the Neolithic, when the Sumerians had not yet invented writing. All that longevity is thanks to some fungi.

Anyway, from 2,600 pounds of yew tree bark, scientists were able to get about 0.35 ounce of what turned out to be pure Taxol. A quick back-of-the envelope calculation shows that to give an adult patient with ovarian cancer a one-day infusion of the drug, you need about 900 pounds of bark. Clearly, you can't give a patient crude bark, or even extracts of bark. To quote Justice Alito, "You'd have to eat a whole [yew] forest to get the value of this." You need to provide an isolated and pure form of the active Taxol. Finding and purifying such active compounds make a worthwhile research endeavor. Giving the searchers and finders some

Figure 11.1. The Crowhurst Yew. Source: Flickr. Licensed under the Creative Commons Attribution-Share Alike 2.0 Generic license. Photographer Donald Macauley.

form of exclusivity to help them bring the drug to market would be a good incentive to keep at it.

In the case of Taxol, the bioprospecting was done and paid for by the NCI, the USDA, and a few universities. It all happened before the mid-1980s, when getting patents on government inventions (or university inventions funded with government money) was, as we saw in chapter 3, not yet the norm. No one received patents on purified Taxol. But having enough Taxol for clinical trials without cutting down the remaining yew trees of the world proved to be the bottleneck. The NCI calculated that for enough Taxol to be available to treat all the ovarian cancer and melanoma cases in the United States would require the annual destruction of 360,000 trees. That's like decimating multiple yew forests per year. Nobody wanted to keep cutting down yews or peeling off their bark.

So, the NCI decided that it had to find a large pharmaceutical company to carry the weight and give it some market exclusivity. It did a deal with Bristol Myers Squibb (BMS), handed BMS the institute's then current stock and supply of yew bark, and gave the company exclusive access to the data so far collected. BMS provided investments to collect further raw material, make Taxol by chemical means, and fund clinical trials. In the end, BMS received a five-year marketing exclusivity thanks to being the first to file for FDA approval.[6] After five years, the powerful drug became, and is now, a generic. It's a shortened generational gift.

Exclusivity in one form or another helped bring Taxol to market. Drug exclusivity does not always have to be due to patents. There are other legal ways of giving someone drug exclusivity for a limited period. Like with Taxol, it may be due to restricted access to supplies, or to data, or to being the first one to get an FDA license. Allowing contemporary companies that invest in the bioprospecting of drugs to obtain some exclusivity on the purified forms of their discoveries is something that society should favor.

That's why I keep bringing up the question: Were you to discover and purify a new Taxol today, could you get a patent on something like "highly purified Taxol"? Is "highly purified Taxol" an article of manufacture even while "isolated *BRCA1* gene" is not? I think so. I think that if properly claimed, purifying Taxol is legally different from isolating a gene.

## A Buffeted Office

For almost two years after *Myriad Genetics* was decided, the Patent Office didn't think so. In March 2014 it issued stern guidelines suggesting that a natural antibiotic is *not* eligible for patents if all you've done is purify it. The office said then that only changing it *structurally* by chemical modification would do the trick, but concentrating it into pure but unmodified form was not enough.[7] Purifying natural antibiotics is just like isolating genes, they said. Sorry, not eligible. Inspired by the suggestion in *Myriad Genetics* that cDNA was eligible because it was a genetic information-carrying *artificial* molecule, the Patent Office said that unless the natural antibiotic was chemically modified, it could not be patented.

Like the Patent Office of 1980 that pushed back against Chakrabarty's wish to get a patent on his live bacteria, the office of 2014 was still being cautious as it pushed back against patenting unmodified purified antibiotics. If you ask Patent Office examiners why they always seem hesitant to expand the boundaries of what is patentable, they will tell you that it is part of their duty to protect the public. They will not grant patents on things that should rightly stay in the public domain. It's almost as though the stranger the invention, the more they dig in their heels. Only tried-and-true inventions seem to keep them comfortable. Electrical circuits or synthetic drugs? Sure. But nothing alive or extracted from nature, please. Ironically, the very essence of the patent system is such that it should reward, not prevent, inventions at the edge of comfort. But the Patent Office is ever cautious.

On the other hand, I feel for the office. It held the line on granting patents on living things until the Supreme Court in 1980 told them in *Chakrabarty* that it should allow anything that is human made. In response, the office allowed thousands of patents on isolated genes until the Supreme Court in 2013 told it that it should not have done so. The poor office has been buffeted back and forth long enough. It is not surprising to see how cautiously examiners step, looking over their shoulders to see what the courts will tell them.

At this point you need to know that no matter how hesitant or daring, what the Patent Office says is not the law. Over and over, federal judges have reminded the Patent Office that it is the Congress and the courts, not the office, that decide what the patent law is and how to interpret it. Several judges have been known to set aside erroneous legal interpretations by the Patent Office. But, until proven otherwise, the office's views do carry a lot of weight among inventors and their lawyers. Sometimes, if the Patent Office digs in its heels, all that's left to do is to take it to court and get a judge to show it the error of its ways. But before going to court, it helps to first try and change examiners' minds in a friendly way. That's what happened with unmodified purified antibiotics.

After the Patent Office published its cautious views that *Myriad Genetics* prohibited the patenting of *unmodified* natural materials, the public criticism was severe. In addition to many of my legal colleagues in the biotech community, my collaborator Chenghua Luo and I published a paper taking the Patent Office to task for its expansive misreading of the *Myriad Genetics* decision.[8] Among several of our arguments, we reminded it of the "Amazon Forest" dialogue the morning of the oral hearing before the Supreme Court. In a remarkable turnaround the Patent Office relented, and a few months later in the same year, 2014, it issued a new interpretation.[9] This one is more relaxed. It no longer says that a patented product needs to be structurally modified from the product in nature. It is now sufficient to demonstrate marked differences in properties or function of the pure material over the natural one.

In our comments to the Patent Office, we also argued that more than absolute purity, what it needs to look at is the *potency* of the drug; that is, How active is it per ounce of formulation? Since Taxol in the bark of a yew is very dilute, its potency per ounce of ground-up bark is extremely low. An ounce of raw bark does not go very far to cure ovarian cancer. On the other hand, when it is concentrated as pure Taxol, its potency is very much higher than in the tree. An ounce of that would go very far indeed.

We explained that since a gene is nothing but information, it is irrelevant how potent it is. The encoded information is the same, regardless of the environment where the gene finds itself. A gene in isolation is no more potent than when it is enmeshed in the genome. In contrast, a highly concentrated anticancer drug like Taxol has a new function and new uses compared to the dilute amounts present in the bark of a tree. It is more potent. Pure Taxol allows an oncologist to administer it in a small manageable injection, as opposed to as part of a tree bark. In purified form it is a markedly different product with new properties absent from the bark. A highly potent new Taxol would indeed be a new "article of manufacture" worthy of patent protection.

While questions and answers at the oral hearing in *Myriad Genetics* are not binding law, the exchanges at the hearing, taken together with the silence in the decision about other natural products, and the change in direction of the United States Patent and Trademark Office, favor the view that *Myriad* seems to be limited to genetics, not other natural products.

A more recent decision from the Court of Appeals, *Natural Alternatives International, Inc. v. Creative Compounds*, further supports this view.[10] The court held that a dietary supplement that contains a concentrated form of the *natural unmodified* amino acid beta-alanine could be eligible for patents if the higher than natural concentration provides a new function.

If I had the chance to rewrite the Pasteur patent claim, I would go from "yeast free from organic germs of disease," to something like "Fast-acting, purified yeast, capable of fermenting beer to completion in fifty percent less time than natural yeast." Such fast-acting yeast is more potent than its germ-laden wild precursor. I am convinced that purified and potent, it is a new product with a new function. It is then an article of manufacture. I believe that the scholarly P. J. Federico, as well as Justices Alito and Ginsburg (if she were still with us), would also see it that way. But if the Patent Office did not, and I were representing a client who is willing and able to fight, I am afraid that I'd have to see the office in court.

In chapter 14 I will show you that in 2014, around the time when the Patent Office was changing its mind on how to interpret *Myriad Genetics*, my legal team managed to get a patent for our client Sanaria, Inc. on genetically unmodified, highly purified mosquitoes that had been made "aseptic." This term is a contemporary synonym for Pasteur's words from the nineteenth century: the modern mosquitoes were also "free from organic germs of disease." The Frenchman would have loved it. His aseptic yeasts and our client's aseptic mosquitoes, as well as unmodified purified natural antibiotics, are perfect examples of articles of manufacture capable of being patented.

# 12

# Enabling Life

The controversies about what is eligible for patents and what is not weren't the only ones flaring up in the first decades of commercial biology. While these were being debated and fought in and out of court, another issue of first impression arose in the 1980s: How does one legally describe the objects of biology so that they can be properly patented? I had a remarkable case that dealt with the issue, and I was lucky that my client was willing to take it all the way to the Court of Appeals.

As with so many cases, this one started with a phone call. At the other end of the line was Marvin Guthrie, the head of the Technology Transfer Office of Massachusetts General Hospital, one of my early clients. Guthrie had first trained as a chemist and then, intellectually restless man that he was, became a licensed clinical social worker. After that, he went to law school. For a few years, in parallel to his private counseling practice, Guthrie served as MGH's attorney in charge of coming down to the emergency room when the ER staff needed a lawyer. The most frequent of such events was when an attending physician spotted a patient who had been physically abused. Often this was a child or spouse with tell-tale bruises that might require legal action. Through his extraordinary combination of psychology and law, Guthrie was the right person to interview the patient or accompanying adult, ask questions, and decide what to do next. He had the calm and empathic temperament of a good therapist and the steely resolution of a tough lawyer. Years later, as he moved to become the head of Technology Transfer at Mass General, these attributes plus his early scientific background made him a much-sought-after hospital resource.

Guthrie had a terrific way to deal with the occasional tempestuous professor who, being at the main teaching hospital of the Harvard Medical School, was used to ordering everyone around. I have met such scientists throughout my career. They invariably believe that they know more IP law than the Court of Appeals and often instruct me what to do. Like a good therapist, Guthrie would listen patiently to such researchers, nod, and say very little. Then, when the storm had passed, he, like a wise lawyer, would remind his interlocutors that while they well may be God's gift to medicine, IP law was IP law. Neither Guthrie nor the scientist could do anything about it. As a result, Guthrie became a highly respected MGH leader. I learned a lot from him.

I recognized Guthrie's calm tone when I picked up the phone. "We have a problem, and I'd like you to handle it," he said. He explained that the case had to do with a cell line called 5D3, which was deposited at the American Type Culture Collection, the ATCC.

My ears perked up. "*In re Argoudelis*!" I thought, "the case on biological escrows!"

## Escrows

Describing life is not simple. Let's say that a scientist has found a microbe in the Badlands of South Dakota. The microbe produces a unique antibiotic. She wants to patent the microbe and the method of making the antibiotic using the microbe as a source. The question is: How does her patent attorney best describe a complicated, live entity so that anyone in the public can get its hands on it and duplicate the production of the antibiotic? While commercial production with the patented bug to make the antibiotic will be prohibited as patent infringement for about twenty years, anyone has the right at any time to study the microbe in the lab and see whether or not it does what the patent says.

The ability to readily reproduce a thing sought to be patented is a central part of the law. This, however, is easier said than done. The description requirements of the patent law are stricter than those that scientists normally use in their published books and papers. The law requires something more than a drawing or an explanatory paragraph; it requires "enablement."

The so-called enablement requirement transcends mere description. It's the idea that to get a patent on the microbe, the inventor needs to make sure that she "enables" the public to reproduce it. Remember the overall deal between society and inventors as conceived in the Venetian Statute of 1474? The deal was that an inventor gets exclusivity in exchange for full disclosure of his invention. Enablement is another word for "full disclosure." At the end of the patent exclusivity period, the public must be able to commercialize the invention without difficulty.

Even earlier, everyone with an interest must also be able to probe into and experiment with the invention although the patent has not yet expired. As mentioned, no one can yet exploit it commercially. But, as the courts have eloquently repeated since the nineteenth century, "an experiment with a patented article for the sole purpose of gratifying a philosophical taste, or curiosity, or for mere amusement is not an infringement of the rights of the patentee."[1] Thus, among the many obligations of the Patent Office when it assesses pending applications is to make sure that the disclosures are "enabling" at the time of examination, at the time of grant, during the close to two decades of exclusivity, and after expiration. Live microbes need to be "enabled," not in the biblical sense, as in " . . . and God enabled Adam with life," but in the legal sense of describing them sufficiently so that they may be reproduced by others. No enablement of the Badlands microbe, no valid patent.

The lawbooks are filled with cases related to the "enablement requirement." Many of them, before the age of commercial biology, dealt with chemical compounds. A chemical compound can be enabled by including in the patent its formula and a recipe for making it. A member of the public reading the patent would see the formula, follow the recipe, and soon have the compound in hand. But a microbe is not a chemical compound. You can't describe a microbe by formula and enable it that way. Not only is it alive, but it is a complex biological structure that you cannot describe with words or a drawing. A microbe does have a taxonomic name, so the reader might know to what genus and species it belongs. But a name is not enough to put anyone in possession of the distinctive bug. There may be many different individual strains of the microscopic Badlands inhabitant that belong to the same genus and species. Yet the older cases in the lawbooks were stricter than that. If extrapolated to microbes, these cases would require that the inventor enable the precise one that produces the antibiotic.

An alternative might be to send everyone to South Dakota, to the spot where our inventor found the microbe. But what are the chances of finding it again? And not just once but as required by law, by every visitor who for many years shows up looking for it? The answer is slim to none. Neither the Patent Office nor the courts accept the concept of sending the public on a wild goose chase to find the darn thing. It's not enough to say, "If you want the microbe, go to the Badlands National Park, dig, and sooner or later you'll find it." The lawbooks make it clear that this is not an acceptable way to meet the enablement requirement. It's too uncertain and requires a lot of work.

So, the attorney in charge of preparing a patent application on this microbe cannot enable it by formula or drawing, or by sending the readers to the Badlands. He must come up with something more creative to make sure the public can replicate it.

The eventual rule on the enablement of microbes arose in 1970, in a remarkable case called *In re Argoudelis.*[2] Being creative, the court combined the law on the enablement of chemicals with the law on escrows. You all know escrows: the financial arrangements used in real estate where an "escrow agent" holds the buyer's deposit money and, once all documents are in order, releases it to the seller at closing. The *Argoudelis* rule is: "Put the microbe in long-term escrow in a laboratory with permanent storage facilities, get the lab to give you a reference number, and include the number in your patent. The laboratory must be bound by contract with the inventor to assure that if anyone from the public asks for the microbe by the reference number, they will get a sample." With this rule, the Badlands bug is enabled.

There are now plenty of "patent microbial depositories" around the world, perhaps none better known to biotech patent attorneys than the

American Type Culture Collection (ATCC), in northern Virginia. The ATCC started off close to a hundred years ago as a repository for valuable biological materials. It still has in its collection deep-frozen samples of Louis Pasteur's original yeast ready to brew beer when thawed. In the 1970s it adapted to the changing law and became a world-class depository for patent purposes.

## Missing Cells

The *Argoudelis* decision ingeniously brought together law and science and came up with a clever escrow solution to "enable" a living microbe so that its discoverer could get a patent on it. By a confluence of events, the phone call from Mass General's Marvin Guthrie gave me the chance to take *Argoudelis* one step forward. It was the mid-1980s, a time when I was still cutting my teeth in the IP law of biotechnology. I wasn't the only one with dental aches. The Patent Office and the courts were also on a steep learning curve. All of us were making new biotech law together, improvising, testing, and arguing with each other, one lawsuit at a time.

After my barely muted excitement at having a possible case dealing with the escrow deposit of microbes, I learned from Guthrie that three scientists at MGH had invented a blood test for the hepatitis B virus. The test's main creativity resided in the use of a distinct subclass of antibodies known as "IgMs." These IgMs were monoclonal antibodies that, as we saw in chapter 2, had been invented by Cesar Milstein and George Köhler in Cambridge, England, less than ten years earlier. The method of producing such antibodies was by extracting them from so-called hybridoma cells, which were a fusion of cancer cells and immune cells. The Mass General inventors believed that the IgM antibodies produced by a specific hybridoma cell line, which they called 5D3, were far better than IgM antibodies from other hybridomas they had in their lab. This line produced superior IgM monoclonals, much preferred for running the blood tests.

One of the three scientists, the late Dr. Jack Wands, was a virologist, hepatitis specialist, and a member of the hospital. The other two inventors had worked with Wands on the invention but had since gone off to form a startup company, Centocor. Their company had an exclusive license from MGH to the test and to the antibodies involved in it.

A patent application on the blood tests had been filed by Centocor's Boston lawyer, David Brook. In accordance with the *Argoudelis* case, one of the MGH scientists, who was now at the startup, had deposited the 5D3 cell line to enable it properly. But having second thoughts, he had since removed it. The deposited cells were too valuable a material, he said, and he did not want the public to get their hands on them so readily. The now-corporate scientist wanted to keep the best cells for Centocor. Brook had been planning to include the cell line deposit by name (5D3) and by its ATCC escrow number (HB-9801) into the pending patent application, but the deposit was now gone. Guthrie wanted to transfer the case to me, as MGH's lawyer, and have me help him solve the problem.

Guthrie was a zealous advocate for Mass General and its academic independence. He explained that the 5D3 cells and the antibodies they produced had been invented at the hospital, they belonged to the hospital, and the hospital was interested in scientific divulgation. The interests of the new startup should be respected, he said, but they were not paramount. He understood that Centocor may prefer secrecy. But MGH did not operate in secrecy. Quite the opposite, its role was to disseminate scientific knowledge. Since the company and MGH did not see eye to eye on what to do about the deposit, and MGH owned the patent application and the 5D3 cell line, Guthrie wanted the case handled by us.

"I'll be happy to help," I said, barely hiding my glee at this first-of-a-kind case. "This is a new one for me. The ATCC deposit was going to be used to enable the 5D3 cells, but the cells may now have become un-enabled. Wow!"

"You sound excited," he mirrored, like the therapist he was. "But this case is bigger than the enablement or not of the 5D3 hybridoma cell line; it also involves a clash between academic and corporate interests. We want to help our researchers' spin-offs succeed, but not to sacrifice MGH's academic values."

"I'm on it," I reassured Guthrie.

After hanging up, I called David Brook, the startup's IP lawyer. He understood *Argoudelis* and, like me, feared that unless we returned the 5D3 cell line to the ATCC, we would have trouble arguing that the cells and their IgM antibodies were enabled. I understood his clients' wishes for secrecy, and he understood my client's policy of open access. While secrecy took precedence for him, free access took precedence for me. But we did agree that enablement of 5D3 was important for obtaining a valid patent. We both put on our thinking caps and tried to come up with a solution that might solve everyone's troubles. We devised a scheme in which we would put the IgM antibodies themselves, but not the cells, on escrow at the ATCC. We would do a deal with the ATCC so that when their escrowed supply of antibodies ran thin, we would replenish them. We were not really convinced that this would legally comply with the enablement requirement of the patent law, but we had ulterior motives.

I set up an interview with John Doll, the director of the Biotechnology Patent Examination Tech Center, crossed the Potomac River to the Patent Office in Crystal City, and proposed the scheme. Doll smiled and acknowledged the thinking that had gone into it but said that the Patent Office would never accept it.

"The ATCC needs to be able to reproduce the cells themselves," he said. "For that, the deposit must be alive and self-replicating, like a cell or a microbe, not inert, like an antibody molecule. And, as an independent escrow, the ATCC cannot depend on MGH. Imagine where this would end if we accepted deposits of inert molecules on the promise that the patent applicant will replenish them when needed." To make sure I understood, he added, "We will reject such a proposal and you'll have to go to court to convince us otherwise. I doubt that the court will side with you." Doll concluded by suggesting that if the MGH wanted a valid patent on the hepatitis test, it had to return the 5D3 cell line to the ATCC. I knew that he was right, and I thanked him for his views.

Brook and I now had what we needed. The rejected proposal served multiple purposes, all of which moved the case forward. For starters, we had put the Patent Office on notice that the best cells had gone missing, a cloud on the ultimate validity of the patent. And, we had received direct warning from director Doll that nothing short of re-deposit would cure the problem. Last, and most important, we had notified the two newly minted corporate scientists that without putting back the 5D3 cells, they might have a shaky patent on which to base the business of their startup. The maneuver did the trick. The scientists agreed that if they received a notification of allowance from the Patent Office, the cells would go back into escrow. Academic freedom succeeded, and Guthrie was pleased. And, I thought, since this ended well, all was well.

But notwithstanding my optimism, the case hadn't really ended. In fact, examination of the Wands patent application had just started. The temporary dis-enablement of the 5D3 cell line was a prelude to a complicated legal battle that ultimately resulted in what is to this date one of the most important cases in biotech patent law, *In re Wands.*

## Broad Scope

By the time Margaret Moskowitz, the examiner in charge of Wands's patent application, picked up the case for examination, the whole issue with the 5D3 cell line had been resolved and no one was the wiser for it. The hepatitis test using the IgM antibodies from 5D3 would again be enabled once Moskowitz allowed the patent claims. But Moskowitz would not grant the claims we had in front of her. Her problem had nothing to do with the enablement of the 5D3 cell line. It had to do with patent scope.

We were not asking for a patent on a hepatitis blood test using only the IgM monoclonal antibodies from the valuable 5D3 hybridoma cells. That would have been a patent of very narrow scope, limited to tests using the specific antibodies extracted from the 5D3 line. There would have been no problem getting examiner Moskowitz to agree to grant us such a patent claim. But a narrow patent blocks only a narrow commercial field. If competitors used their own IgM monoclonal antibodies, different from the ones from the 5D3 cells, they would be able to use the inventive test without infringing the patent. In the jargon, it would be easy for competitors to "design around" such narrow patent claims.

Imagine that you have invented an ingenious, Siri-like "mechanism X," which allows a pivoting chair to turn on a verbal command. All you need to say is "Chair, turn 90 degrees," and the chair will do your bidding. You build a prototype in the form of a dining room chair. You could file for and get a patent claim on "A dining room chair with mechanism X." This is known as a "specific" claim. A specific claim, however, would protect just dining room chairs, and that is of limited worth, legally and commercially. Anyone can copy mechanism X, the main invention, and put it into all kinds of other chairs—such as armchairs, desk chairs, living room chairs, or patio chairs—without infringing your specific claim. The kind of chair that carries mechanism X is at best a sideshow.

What you really want is to get a broad patent claim that says, "*Any chair* with mechanism X." Such a claim is known in the jargon as a "genus" claim. It is of a scope broad enough to cover the entire family of chairs, not just those around a dining room table. In our case, "mechanism X" represents the hepatitis test using IgM antibodies, which is the main invention. The "dining room chair" represents the specific IgM antibodies from the 5D3 cell line. These may be the best ones for the test but are a sideshow.

Mass General and its license holder Centocor wanted a patent with a much broader scope than one limited to the IgM antibodies from 5D3. The ideal patent would be one that blocked commercial use of the hepatitis blood test no matter what IgM antibodies Centocor's competitors used. We aimed for a genus patent claim that extended to *all* IgM monoclonal antibodies against hepatitis, not just those from 5D3. But Examiner Moskowitz was not willing to give us a claim of such breadth. Moskowitz's main argument was that a deposit of the 5D3 cells enabled specific patent claims to a test using antibodies from the 5D3 cells and nothing else. Essentially, she was saying, "You can have mechanism X, but only on dining room chairs, not on all chairs." I knew that she was right.

So, I changed tack. I openly agreed with Moskowitz that a deposit of 5D3 could not enable the full scope of a broad genus claim to a test using all IgM monoclonal antibodies. But I also said that we would not deposit at the ATCC any more cells than the 5D3 ones. My main argument to examiner Moskowitz was that when one deals with patent claims of broad scope, deposits are unnecessary. Forget *Argoudelis*. A claim to a blood test of broad scope can be enabled by routine, state-of-the-art screening, no biological deposit needed. The core of my pleading was that all one had to do was consistently generate a pool of monoclonal antibodies against the hepatitis virus, select those of the IgM subclass, and screen again for those that bound strongly to the virus. You would then have an unlimited supply of useful antibodies. "Routine screening" was my cri de coeur. "More deposits" was Moskowitz's.

My new take on the case was quite an about-face. After dealing with secrecy-driven corporate scientists who didn't want a deposit at all, after meeting with the director of the Biotech Group to get his views, and finally convincing the scientists to replace the deposit, I was now arguing that such a deposit was not even necessary! It is always important to remain flexible when lawyering a case.

Examiner Moskowitz and I went back and forth for a year of written objections and responses. There was no solution to our disagreement. So, my final paper to Moskowitz included notification that I would take the case to appeal. I diplomatically explained to her that, with all due respect, we seemed to have reached an unfortunate impasse, and only higher judicial review could resolve the issue. Surprisingly, she welcomed the appeal. She said that it would help her, and other examiners, clarify the role of individual biological deposits when dealing with patent claims of broad scope, claims that encompassed many more than the specifically deposited biological materials. The biotech examining corps had been waiting for a court ruling, and this was the case. After an intermediate appeal to the Patent Office Board of Appeals, which affirmed Moskowitz, the case finally reached the Court of Appeals for the Federal Circuit (CAFC).

## Quicksand

The CAFC is housed in a beautiful modern red brick building facing Lafayette Square, in downtown Washington, DC. The Court is kitty-corner from the White House, so, when in 1988, on the date of the oral hearing in the *Wands* case, we climbed the steps leading to the main entrance, we had the White House to our right and the yellow façade of St. John's, the

"President's Church," behind and to our left. It was a crisp spring morning, with some of the square's cherry trees still in bloom.

It was my first appeal. I was nervous as could be as my legal team and I walked into the courtroom and sat at counsel's table. It didn't help that sitting next to me was a gloomy young associate from our firm who was convinced that we would lose. Ignoring his pessimism, I walked up to the lectern, took a deep breath, and started.

"Good morning and may it please the court. My name is Jorge Goldstein, and I am counsel for Party Wands," I said. "I will take fifteen minutes and reserve five for rebuttal," I added, as is customary. The three judges looked at me and didn't even nod.

But the judges soon perked up. For the next quarter of an hour, I sparred with them about enablement of biotech patent claims, biological deposits, the reach of the *Argoudelis* precedent, and why it didn't apply here. My argument was that routine screening was all that was necessary for Wands and his coinventors to get a genus claim to a blood test, one that couldn't easily be designed around.

"So, Mr. Goldstein," interrupted Judge Edward Smith, "Why would you make a deposit of the 5D3 cell line, if you tell us now that you don't need a deposit?"

That was a delicate question, because it went to the apparent contradiction of not needing a deposit yet being willing to deposit 5D3. But I was ready for it. I had rehearsed the answer a week earlier, during moot hearings. My answer had to do with another portion of the patent statute, this one dealing with the so-called best mode. The best mode requirement obliges an applicant filing for a patent to reveal the best cells, methods, reagents, or materials to carry out his invention. It prevents an inventor from concealing from the public his most preferred incarnations. The law at the time of the *Wands* case was that if years later, during infringement litigation, it was discovered that the "best mode" was withheld, the court might well invalidate the patent.

"We deposited the best cells we had at the filing date to comply with the best mode requirement," I answered without hesitation. "But, Judge Smith," I continued, "the deposit of any one cell line isn't needed to comply with the enablement requirement for the broad testing claims we are seeking. Routine screening will do for that." The judge nodded at my answer, and the moment passed.

Arguing for the Patent Office was John Raubitschek, the associate solicitor for the commissioner of patents. Raubitschek got up, went through the formalities, and started his argument. He mirrored examiner Moskowitz's view that deposit of one specific cell line was insufficient to enable broad genus claims. As he was ready to expand on this theme, Judge Pauline Newman interrupted him. Newman, whom we met in chapter 5, is another PhD scientist-turned-judge. She was then, and to this day remains, known for her finely tuned and carefully dissected dissents. I held my breath waiting for her query.

"Mr. Raubitschek," she asked, "if one deposit is insufficient to enable broad claims, how many would you recommend?"

I loved her question. There was no precise answer. The uncertainty was the very reason I had decided to drop the idea that biological deposits were necessary to enable patent claims of broad scope. A genus claim, like ours, covered a very large, and ultimately indeterminate number of different individual antibodies. So, would we have to deposit thousands of different cell lines, each generating a different IgM antibody? Who could tell if even thousands of deposits would be enough? The answer was that no one could really tell. Newman's line of questioning led to quicksand.

I couldn't wait to hear what Raubitschek had to say. I had told my skeptical associate that if the court asked how many cell lines had to be deposited to enable broad claims, the Patent Office associate solicitor would head into the swamp. Raubitschek stepped right in.

"Well, Your Honor," he said. "It depends."

"It depends on what?" asked Newman, her eyes sharpening. "Would you say a hundred deposits would do? A thousand? How many?"

"Each case is different," he hemmed and hawed.

"OK, but what about in *this* case?" asked Judge Newman. "Doesn't the Patent Office want some clear guidance from us and not just a decision that says, 'It depends?'" she added, twisting the dagger.

She finally let him off the hook. Raubitschek finished his argument the best he could and sat down.

Before walking to the lectern for my five-minute rebuttal, I whispered to my associate, "We're going to win this one." The young man finally smiled.

I looked at Judge Newman, and said, "Judge, you asked the right question." And, I added, "As the Associate Solicitor showed us, there is no answer to the question, 'How many deposits are needed to enable a broad claim?' The answer is no deposits are needed if all that is required is routine, state-of-the-art screening. That's how one enables broad patent claims in biology." And I sat down, having used up at most one minute of rebuttal.

A few months later, the court ruled in our favor. Judge Smith, who had asked me why we would deposit even one cell line, wrote the opinion. The court held that no deposits are needed to enable broad genus claims to a blood test if all that's necessary is "routine experimentation."[3] The decision was 2-1 with none other than Judge Newman, she who had given Raubitschek such a hard time, dissenting. Her complaint was that there was not enough evidence as to what was routine and what was not routine experimentation in the field of making IgM monoclonal antibodies.

But the two other judges sided with Mass General. The *Wands* case established a framework for how to legally analyze whether broad claims are enabled or not. The framework, which, relying on earlier precedent I had proposed to the court, looks at eight questions. These questions have since become known as the "*Wands* factors." They are: How much experimentation is needed to enable the full scope of a claim? How much guidance is present in the patent specification? Are there any actual working examples in there? What is the nature of the invention? What is the state of the art? How much skill do others have in this field? Are experimental results in this field generally predictable or unpredictable? Finally, how broad are the claims? Note that not one of these questions asks about biological deposits. The court asked and answered the eight questions in our case. It weighed the plusses against the minuses and concluded that, on balance, our broad claims were enabled.

This legal framework has withstood the passage of almost thirty-seven years. *In re Wands* has since been cited as binding precedent in more than 1,200 court opinions. Even when the courts distinguish the actual facts from those in my 1988 case, they continue doing so based on the eight *Wands* factors. In 2023, and for the first time since *Wands* was decided, the Supreme Court, in *Amgen v. Sanofi*,[4] did precisely that. The case pitted Amgen's broad patent on cholesterol-decreasing antibodies against Sanofi's complaint that Amgen's patent was far broader than deserved. "All they made is a few antibodies and they want a patent for thousands and thousands," objected Sanofi. The Supreme Court framed the central issue as: Was Amgen's contribution to the public as broad as the patent they received?

There were amicus briefs from close to two dozen friends of the Court, representing the who's who in biopharmaceutical research. These friends were on both sides of the question: some (Glaxo, Bristol Myers, AstraZeneca) supported Amgen, and others (Eli Lilly, Pfizer, Genentech) sided with Sanofi. Regardless of which side they took, none of the amici undermined *In re Wands*. They all agreed that the eight factors from 1988 were the proper way to look at things. Their differences were whether, even while using the *Wands* analysis, *in this particular case* Amgen had obtained broader claims than warranted.

In analyzing the *Amgen* case, the Supreme Court picked up on Judge Newman's dissent in 1988, when she had asked how much experimentation was needed to enable the full scope of Jack Wands's patent.[5] The analysis, said the Supremes, is all about weighing inventive contribution and patent scope. The more of one, the more of the other. The Court explained that if the amount of experimentation necessary to enable the full scope of a broad patent claim is similar to what it took the inventor to make the invention in the first place, it is "undue." Such was the case with Amgen's large family of antibodies, the Court concluded. Amgen is entitled to patent one or a few antibodies but not thousands. Since Amgen's patent was broader than warranted, it was invalidated. Yet in deciding against Amgen, the Court did not overrule or even criticize the *Wands* case; through its reasoning it implied that the Court of Appeals had used the *Wands* factors properly. (And of course, I sighed a sigh of relief.) My dear *Wands* factors survived. I am pleased to say that they remain as much a part of biotech patent law as any case on the books.[6] The escrow rule of

*Argoudelis*, which so stimulated my early thinking, is still good patent law when applied to narrow claims drawn to specific biological individuals. However, when dealing with broader claims, the analysis is based on the 1988 *Wands* case, the one that allowed me to leave my mark on the law.

Off and on over the intervening decades and until his untimely death in 2023, I chatted with Jack Wands. He was the director of the Division of Gastroenterology/Hepatology and of the Liver Research Center at Brown University Medical School. Given the eminence of the legal precedent that bears his moniker, his has become a famous name in the law of biotech patents. Jokingly, I sometimes called him "In re" instead of Jack, and he always laughed.

Even though his patent on the hepatitis test expired in 2007, the 5D3 cell line is still listed in the ATCC catalogue, available to everyone. If you want a sample of this once controversial cell line, contact the ATCC, send it $249.00, and ask for the cell sample by its escrow number, HB-9801. You can then use it for any purpose you wish, whether it be commercial or for "gratifying your philosophical tastes."

# Part III

# From Microbes to Mammals

I have some colorful stories to tell you of patent cases in the world of commercial biology. They came after the dust had settled on many of the "issues of first impression." As the years advanced, we, inventors, and attorneys, had clearer direction from the courts on what was or was not patentable and how best to describe it. And based on that guidance, we engaged in legal fights involving inventions across the biological universe of artificially made living things: from small microbes, to insects, to plants, to large mammals. And as equally vivid as the inventions at stake, were the protagonists involved.

# 13

# Microbes

One of the most colorful patent cases I have been involved with had to do with the pigments that give salmon their attractive orange-red hues. The fish, however, were not the only colorful creatures in this tale. The humans were not far behind.

When salmon are born in their home rivers and swim out into the ocean, they are silvery white. They only turn orange as they are about to reenter their ancestral rivers to spawn. Figure 13.1. shows two sockeye salmons, one in its so-called ocean-phase, and the other in its so-called spawning-phase. The two drawings are not of the identical individual taken before and after approaching its home river. Getting a photo of one such individual would be extremely hard. Yet the average side-by-side differences are dramatic enough. Not only is the fish on the right darker (imagine an intense orange red) than the silvery one on the left, but it has also changed its shape, looking more streamlined. This is probably in preparation for the upcoming hard labor of swimming upstream in strong mountain currents and to then procreate.

The fish do not produce the red-orange color themselves; they take it in after they eat colorful yeasts floating alongside their ocean routes back to their home river. The yeasts, *Phaffia rhodozyma*, in turn, contain large amounts of carotenoids, the same chemicals that give carrots, flamingos, and autumn leaves their lovely hues. Now, the thing to understand is that when salmon are grown in fish farms along the coasts of Norway, Alaska, or Chile, they are also white, unless they are fed *Phaffia* yeasts. There is a competitive market in selling *Phaffia* yeasts to fish farmers, who then supplement their feed with the colorful bugs. The more color the yeast

Figure 13.1. Sockeye salmon (*Oncorhynchus nerka*) in ocean and spawning phases. *Left side*, ocean phase. Source: Wikimedia Commons. Author US government. Taken from page 5 of US Government Printing Office Pamphlet 1996-792-501: Lake Washington Ship Canal Fish Ladder (1996); *right side*, spawning phase. Source: Wikimedia Commons. Author A. Hoen and Co. Scanned from plates in Barton Warren Evermann, and Edmund Lee Warren, *The Fishes of Alaska*, Department of Commerce and Labor Bureau of Fisheries (Washington, DC: 1907). Both sides of the figure are in the public domain.

cells make, the better. A farmer spends less money on yeast if the yeast produces more orange per gram of feed.

And, of course, if there is competition for improving yeasts, there will be biotech patents. Archer-Daniels-Midland (ADM), a giant multinational company from Decatur, Illinois, came to me in 1997 to help it file a lawsuit dealing with the vibrant red tone of supermarket salmon. ADM had patents on a method of fermenting *Phaffia* yeasts so that they would produce high quantities of the much-prized red-orange carotenoid. You may remember from chapter 9 the major 1980 decision of the Supreme Court in *Diamond v. Chakrabarty*, which held that aliveness was no impediment to patentability. The ADM patent also included protection for live *Phaffia* yeasts in all their ruby brilliance. The patented yeasts had the ability to make unnaturally large amounts of the orange-red color, allegedly producing more than 3,000 parts per million of the colorful stuff per gram of dry yeast.[1]

## Red Yeasts

Starting in the early 1980s, the growing biotech industry steadily came up to speed on the reprogramming of living things. And we patent attorneys were not far behind our creative clients, learning how to safeguard and fight for their biological inventions. We also had to study some complicated genetics. Yet no matter how much we studied, sometimes our subjects were impenetrable. That was the case with the ADM litigation over their unusual—and very complicated—yeasts.

Creating the genetically engineered microbes that produced insulin at Genentech involved using simple bacteria. One of the most useful bacteria for such purposes is *Escherichia coli*, the non-pathogenic bug we all carry in our guts. Most *E. coli* bacteria are harmless, and they reproduce faster than rabbits; actually, much, much faster. They do so by splitting into two cells every twenty minutes, an exponential growth that slows only when they run out of food or get too crowded. And because they are non-pathogenic and divide quickly, *E. coli* bacteria have been the darlings of the biotech industry.

But bacterial cells are only one of many different microbes used in commercial biology. Bacteria are members of one of the two major empires into which all life is split; they are "prokaryotes." Prokaryotes lack an enclosed nucleus; their genome free-floats inside their unicellular selves. The other empire of cells is that of "eukaryotes." Eukaryotic cells are about ten times bigger than their more primitive ancestors like *E. coli*. Reflecting their complexity, eukaryotes have much more DNA than prokaryotes. Their genomes are not free floating like that of prokaryotes but are encircled inside a balloon-like nucleus. Pasteur's beer brewing yeasts and the red *Phaffia* yeasts are eukaryotic cells. And, when such eukaryotic cells aggregate to form tissues, they become full-fledged "eukaryotic organisms," such as mosquitoes, plants, or animals. We humans are eukaryotic organisms. If you wish to see a human eukaryotic cell with an enclosed nucleus, go back to figure 10.1 in chapter 10, where I show such a cell and the DNA strands of its unwinding genome.

Eukaryotic *Phaffia* cells are more complicated than prokaryotic bacteria. Just how complicated was made clear to me in the ADM patent case. The litigation was between Igene Biotechnology and ADM. The lawsuit started when ADM sued Igene, a (very) tiny company in Columbia, Maryland, for patent infringement. When Igene responded to the complaint a month later, they countersued ADM for stealing its proprietary yeast cells. The lawsuit soon became a David-against-Goliath battle, and I represented the Philistine giant, not the underdog.

The case was technically complicated even for me; and I don't usually become intimidated by dense science. A *Phaffia* yeast cell is about as difficult a microorganism as I have ever had to deal with in litigation. *Phaffia* has somewhere between nine and eleven chromosomes, to *E. coli*'s none and to humans' forty-six. *Phaffia*'s genome is five times longer than that of *E. coli*, meaning that it encodes five times more genetic information than the simple bacterium.[2] The genetics of *Phaffia* are not easy to understand, not to mention describing them to a judge or a jury, especially to a jury composed of people with no scientific training or interest.

In 1997 I was not a trial lawyer; in fact, I still am not, and, by choice, never will be. My specialty was—and to this day remains—a certain crossover ability to bridge complicated biology with abstruse areas of patent law and explain them to laypeople. Whenever possible I have avoided the high-stress lives of trial litigators, even though they are oftentimes the darlings of the news media.

ADM had earlier retained me and my firm to get the patent on the improved red yeasts. After the patent issued, the company asked us to help with the lawsuit against Igene. It was hoping that I could bridge the knowledge gap between the judge or jury, and the scientists and their complex *Phaffia* bugs. Just in case we would ever be before a jury, I brought in David Cornwell, a partner who was good in court and carried his smarts very modestly. Our firm's David would defend Goliath.

Since the case was about improved ways of coloring salmon, I was immediately curious to find out if the color made any difference in the flavor of the fish. I have made it a habit over decades of legal practice never to eat my clients' patented foods. I have been offered ingenious coffee replacements made from hickory, extruded chicken mini-cylinders, colorfully striped synthetic crab legs, and other assorted novel and inventive edibles that I have helped patent. Neither fear for my health, nor ecological concerns, nor monopolies are the true reasons why I don't eat my clients' comestible inventions. The problem is that I don't always believe my enthusiastic clients when they tell me, "Try this, it tastes just like the real thing!" I am not a good liar, and my face would just give me away after the first bite. I'd rather keep the trust of my clients than pretend that I like their ersatz foods.

Curious, however, I did a side-by-side taste test comparing the meat of two Pacific Coho steaks having different colors. I should add that I also avoid farm-raised salmon unless I know that they're grown under sanitary conditions. So, I compared the meat of Coho caught during their life while still in the ocean, a time when they are silvery, with the meat of mature freshwater adults, which have a bright orange color. Both, seasoned with some dill-spiked mustard, were equally excellent. I suspect that just as the colorful red-chested robins that strut around in my backyard in the spring are trying to attract females, the red in the Coho males has more to do with sex than savor. But regardless of the similarities in taste between ocean and spawning salmon, American consumers want their salmon to be a bright red-orange hue. So, coloring farm-raised fish by feeding them high-yielding *Phaffia* yeasts is an industry all by itself.

## David versus Goliath

Back to ADM versus Igene. In July of 1997, our firm charged Igene in Federal District Court in Maryland with commercializing high-yield *Phaffia* yeasts that were protected by ADM's patent. It turned out that ADM and Igene had been negotiating for a while, trying to reach some sort of agreement in which Igene would provide patented strains of the *Phaffia*, and ADM would then commercialize them with their networks and marketing savvy. This is a classic arrangement in biotechnology between small research companies and large multinational distributors, each bringing to the table their own talents. In this case, the situation seemed reversed. It was ADM who had the patented red yeasts, not Igene.

ADM, founded in 1902 by George Archer and John Daniels, started as Archer-Daniels Linseed Co. and, after buying Midland Linseed in the 1920s, became Archer-Daniels-Midland. It is today involved in oilseed and corn processing but runs one of the world's largest networks of grain elevators, transportation, port, trains, and storage facilities. All of these have to do with the production and selling of food and its ingredients. It is a huge and powerful company. I remember flying into Decatur, about an hour from Chicago by puddle hopper, and seeing ADM's grain elevators and barge-loading docks rising from the flat Midwestern lands like giant mirages in the shimmering hot air. The company seemed to own the prairie.

A month after we filed the lawsuit, we received Igene's defense and counterattack. It accused ADM of stealing its trade secrets. Its allegations were that during the negotiations, ADM had somehow pilfered *Phaffia* strains and other secrets of Igene. ADM was adamant that no such thing had happened and made a case that in fact it had been Igene who had stolen ADM's yeast strains.

At the time, we were not unduly surprised by Igene's allegations, because ADM was a tempting target for accusations of inappropriate behavior. Its past legal difficulties had included several investigations by the Department of Justice. The most notorious of these was the 1993 charge that ADM had engaged in price fixing in the international lysine market. The company was fined $100 million dollars as an antitrust penalty, and two of its top officials, Michael Andreas, former vice chairman, and Terrance Wilson, the retired chief of its corn processing division, were sentenced in 1999 to two years in federal prison. In the wake of these events, we were not taken aback that Igene was accusing ADM of acting deviously. It's not that I believed Igene; I was not shocked that Igene was performing the poor and innocent David to our rich and sullied Goliath. The smaller company probably figured that this strategy might play well in court.

The fight between the two companies quickly became focused on the identity of the yeast cells. Were the cells in the hands of Igene identical or similar to the ones ADM had patented? Did ADM have any cells that were identical or similar to the ones that Igene was alleging had been stolen? What was bewildering to all of us was how complicated it would be to answer these questions. I thought that running a straightforward genetic fingerprint of Igene's cells and comparing it to the fingerprint of ADM's patented ones would be simple. Not even close. We hired top expert geneticists. Igene did too. I came to understand that we might have to prove that the yeast cells in Igene's hands were a *progeny* of the original ones it might have had—and perhaps vice versa. The question became: How do you determine that a yeast cell is a daughter or a granddaughter, or even a descendant further down the line, of the original ancestor? The experts on both sides scratched their heads. We got into such a disagreement about what to test and how to test it that Judge William Quarles Jr., who was presiding over the litigation, got himself his own expert.

The lawsuit turned out to be nothing less than a full-fledged paternity fight. The yeast cells played the role of the children, or perhaps the great-great-grandchildren. Either ADM or Igene, or both, played the role of the quarreling parents. It was a genetic and legal mess.

## If the Glove Doesn't Fit . . .

The back-and-forth on how to test went on for almost two years, at which point we got the biggest surprise of them all. In February of 1999, we received notice that Igene had changed lawyers and that none other than Johnnie Cochran—yes, the one of O. J. Simpson's fame—was entering an appearance. I shook my head in disbelief. Here was one of the most complicated genetics cases I have ever litigated, and suddenly Cochran was my opponent. "The man may know nothing about yeast genetics," I thought, "but he sure knows about juries. And, since he is known as the champion of the underdog, he surely knows about defending Davids against Goliaths, especially when this David's recovery might be close to three hundred million dollars."

Let me tell you about juries in patent litigation. Believe it or not, any of the parties to a patent lawsuit in the United States has the constitutional right to a jury trial. Good trial lawyers may not know much about the genetics of *Phaffia*, but they sure know how to tell a good story, one filled with colorful metaphors, which pits their good client against the evil opponent—and vice versa. In a patent battle before a jury, the scientific truth gets fogged up quickly. Often, it turns into a clash about whose expert witness is more credible or whose lawyer seems more trustworthy.

Many a juror interviewed after a verdict for the defendant (or plaintiff) has been known to say, "I didn't understand any of that stuff about recombinant DNA, or fusion proteins, or monoclonal antibodies, or whatever that was. But I sure didn't like that arrogant lawyer from the plaintiff's (or the defendant's) side, who thought he was smarter than everyone in the courtroom, and never stopped trying to show it."

Cochran was famous from his 1995 win in the criminal defense of O. J. against the state of California. One of the highlights of that case was the back-and-forth between the prosecution and Cochran's defense team about the significance of DNA evidence. DNA fingerprinting of O. J.'s blood had become a central theme, with Cochran ultimately convincing the jury that the evidence was tainted and unreliable. We guessed that Igene had hired him because of his masterful command of juries and his recent court experience with DNA. I have always thought that if the facts in a jury case are very complicated, as with the paternity of *Phaffia* yeasts, it pays for a party to bring in an over-the-top lawyer, well versed in the arts of obfuscation.

We were thrilled that such a unique opponent was sitting at Igene's defense table, but we were also worried. We were not sure how to prepare against the expected theatrics. We were sure that Cochran would show . . . shock! . . . even outrage! . . . as he accused our client, the powerful ADM, of stealing valuable yeast strains from his weak and innocent startup. One member of our team even composed a poetic reference for Cochran. It was inspired by Cochran's famous summation in the O. J. case when, talking about the ill-fitting glove into which his client unsuccessfully tried to put his hand, he said to the jury: "If the glove doesn't fit, you must acquit." You may recall, of course, that the glove didn't fit, and the jury did acquit. We referred to the celebrity lawyer as, "Johnnie-let-go-of-the-*Phaffia*-or-we'll-call-the-mafia." It didn't have much meaning, but the rhyme was perfect, and it invariably brought a laugh to our team.

By mid-1999 a bewildered Judge Quarles ordered both sides to try mediation and put further yeast testing on hold. By late that year we had achieved nothing but continued bickering. By early 2000 Dr. Nicholas Ambulos, who was the judge's expert and a professor of microbiology at the University of Maryland School of Medicine, was asked to renew his microbial paternity work. Disputes about how to test continued for two more years. Then, in July 2003, the exasperated judge ordered that we go to trial with whatever testing we had.

There was never a trial. In an anticlimactic ending, the *Phaffia* case settled late in 2003 without a final determination of the yeast's ancestry. Both sides gave up. Maybe everyone just became exhausted from all the back-and-forth about genetic testing of yeasts. Maybe not seeing a clear outcome that favored it, each side figured that settling was better than playing the odds. Or maybe it was that in December 2003 Johnnie Cochran was diagnosed with a brain tumor. Igene may have felt that without its jury genius its chances of winning had gone down. We never saw Cochran do his magic. He dropped out of the case and, sadly, passed away two years later.

In 2009 ADM and Igene formed a joint venture named Naturxan to make and sell improved *Phaffia* strains for use in the fish farming industry.[3] The negotiations they had started twelve years earlier came to fruition after the lawsuit. It took a rancorous set of accusations and charges, huge legal and expert fees, and, for all of us, a brush with celebrity law, to get these two companies to finally agree.

If it weren't for the appearance of Cochran, this lawsuit may not have stayed in my memory. Most patent disputes never get to litigation; they are resolved amicably with licenses or, as in this case, the formation of joint ventures. And of those that do get into the courtroom, only a few go the full nine yards. They only reach the bitter end when one large, well-funded company, say a plaintiff holder of a patent, wants the defendant, its competitor, usually another large, well-funded company, out of the market. If you want somebody out of the market, you don't offer a license, you get a permanent injunction in court at the end of a costly trial and get it affirmed on appeal. It is surprising that ADM and Igene fought as long as they did in the *Phaffia* case. After all that complicated, and ultimately unresolved, strife, they got together and formed Naturxan. They could have saved themselves a lot of money and aggravation.

I have learned that clients don't always act in entirely rational ways.

Sometimes there is bad blood between competitors, carried over from past clashes. Sometimes the issue becomes personal, as when one thinks that the other has stolen something. And, sometimes, a lawsuit is just a way to gain leverage to negotiate a better deal. "War is merely the continuation of politics by other means," as Carl von Clausewitz is meant to have said.

# 14

# Mosquitoes

Let's now go from patenting small microbes to patenting midsized insects, such as mosquitoes. Mosquitoes have been the bane of humanity for all of history. And that's not only because they are a whiny nuisance or cause itchy welts after biting. They can also be deadly, as in malaria deadly.

To add human perspective to the IP story of fighting malaria, let me introduce Dr. Stephen Hoffman, its hero. In an interesting twist of scientific fate, Dr. Hoffman was also involved in early work that led to the messenger RNA (mRNA) vaccines used in 2020 to effectively fight the COVID-19 pandemic. Let's start with vaccines and then move on to mosquitoes.

## Nucleic Acid Vaccines

The first time I met Dr. Hoffman, in the mid-1990s, he was the director of the Malaria Program at the Naval Medical Research Center, in Bethesda, Maryland. He was recommended to me by another client, Vical, Inc., of San Diego, California. Vical told me that Hoffman was someone who had recently succeeded in using DNA vaccines, a revolutionary new technology that Dr. Phillip Felgner, together with a team at Vical and their academic collaborators, had developed.[1] The Felgner team's invention of injecting nucleic acids into a live animal and generating immunity happened in 1989.

Felgner's inventions were way before the appearance of BioNTech in Germany or Moderna in Massachusetts, the two companies that are the creators of modern mRNA vaccines against COVID-19. During the 2019–21 pandemic, BioNTech and Moderna showed that you could inject mRNA encoding the COVID-19 viral spike protein together with modern-day nanolipids and bring about a highly efficacious vaccine against the pandemic.[2] But the basic concept of so-called genetic vaccines was invented by the Felgner team and its academic colleagues. Our firm filed plenty of patent applications to protect the Felgner discoveries of nucleic acid injections.

The Vical folks, however, had run into trouble at the Patent Office, which kept rejecting their inventions of nucleic acid vaccines as simply not credible. Without additional proof, said the Patent Office, injecting straight DNA or mRNA and getting immunity were just plain unbelievable. Vical thought that perhaps Dr. Hoffman could help them change the office's mind and get some patents issued.

## Navy Whites

That's where Dr. Hoffman came into my picture. I retained him as an expert in the field of vaccines, who could explain to the examiner that he had used DNA vaccines for malaria and how real the results were.[3] He sure had the credentials. His online bio shows him as having an alphabet soup of titles and honorifics: "MD, DTMH, DSc (Hon), FASTMH, FIDSA, FAAAS, FAAM, CAPT, MC, and USN (RET)." You can look them up if you're curious to find out what they stand for. As far as I was concerned, I liked him as an expert because he was a world authority on tropical medicine and infectious diseases. He was a captain in the US Navy, to boot.

I visited him in his then cramped lab quarters in Bethesda, where we discussed the outline of the testimony that he would give to the patent examiner. He asked me if he should wear his whites to the interview, and I said, "Sure," figuring that a naval uniform couldn't hurt Vical's case. Little did I know.

On the day of the interview, I invited Dr. Hoffman to my office before going together to the Patent Office. I received a call from our receptionist, who announced that he had arrived and was in the waiting area. I walked over to the reception area and shook his hand. I knew something was up when I saw a few more lady secretaries than usual standing around the main desk, buzzing and smiling. I took a second look at Dr. Hoffman. He was the perfect model of a modern navy doctor: tall, green eyes, light hair, white uniform, including several decorations. He was wearing his captain's white-and-black combination cap with gold buttons, an officer's crest device, and a silver shield over crossed golden anchors, an eagle on top. He had a disarming and boyish smile, straight out of central casting for a handsome navy man.

We drove over to the Patent Office. I've never seen an examiner agree so quickly to drop all rejections in the case and allow the Vical application. She was all admiring smiles and nods during the hour or so during which Dr. Hoffman, his black, white, and gold cap staring at her from the top of the desk, explained how injecting DNA and mRNA into human muscles worked very well and quickly generated vaccines. I think that she believed him even before he uttered the first word.

We obtained several patents for Vical based on Dr. Hoffman's expertise.[4] All of them expired around 2018, so by the time the COVID-19 pandemic hit the world two years later, they were no longer in force. It pays to be ahead of conventional thinking, but if you are way ahead, your patents may expire before the rest of the world catches on.

I am convinced that Dr. Hoffman's scientific authority as a vaccine expert helped Dr. Felgner and Vical get their patents. I am also pretty sure that his good looks and his navy whites did not hurt. Lawyering a case takes many forms.

## Femmes Fatales

A few years after these events, in 2002, Dr. Hoffman left the navy, founded a company called Sanaria, Inc., and asked me to be his lawyer. His goal was to develop a vaccine for malaria. The name of his company contrasts with the Italian name for the disease, which he was intent in eradicating once and for all. The disease name comes from "mal' aria," bad air; his company's name is "San' aria," healthy air.

Malaria is a plague of the tropical world, where mosquitoes thrive. The female of the Anopheles mosquito carries in her gut a parasite, named *Plasmodium falciparum*, which causes malaria. She feeds at night and, before she sucks up blood, injects pieces of the parasite, called sporozoites, into the hapless recipient. The live sporozoites go into the victim's liver, where they grow and morph, and eventually break free and infect blood cells, which then start circulating. When the next hungry bug lands to eat, the parasite goes from the blood circulation into the mosquito, where it develops into more sporozoites, and so on. *P. falciparum* is a nasty parasite that develops partly in mosquitoes and partly in humans. It's not the mosquito that causes malaria; it only carries the pathogen that does and is therefore known as the "vector" for the disease. Anywhere from a week to a month after the injection of sporozoites by the Anophelene femme fatale, the infected victim develops headaches, fever, joint pains, vomiting, and worse. In serious cases, malaria can cause seizures, coma, or death. There are no easy cures for severe cases, and mortality can reach close to 30 percent. About 620,000 people died of malaria in the world in 2021.[5]

Because of the complexity of the parasite's two lifecycles, it has been impossible to develop a highly effective vaccine. That is, until Dr. Hoffman, the chief brains behind my client Sanaria, arrived on the scene. Treating or preventing malaria has been a very hard nut to crack, but Hoffman has made it his lifelong quest. There are all sorts of troubles: drug resistance of the parasites, insecticide resistance of the mosquitoes, and the lack of a good vaccine. Bed netting is about as good a solution as there is. Ideas have been floated to genetically engineer mosquitoes to make them unproductive vectors. The results suggest that such mosquitoes in lab boxes can indeed spread resistance to the malaria-causing parasite by thriving and mating with non-modified mosquitoes. It is not clear, however, if these results will hold up in the wild.[6] Not surprisingly, the World Health Organization has echoed the thinking of many when it expressed concerns about releasing genetically engineered mosquitoes into the environment.[7]

As far as vaccines, the story has not been all that promising either;

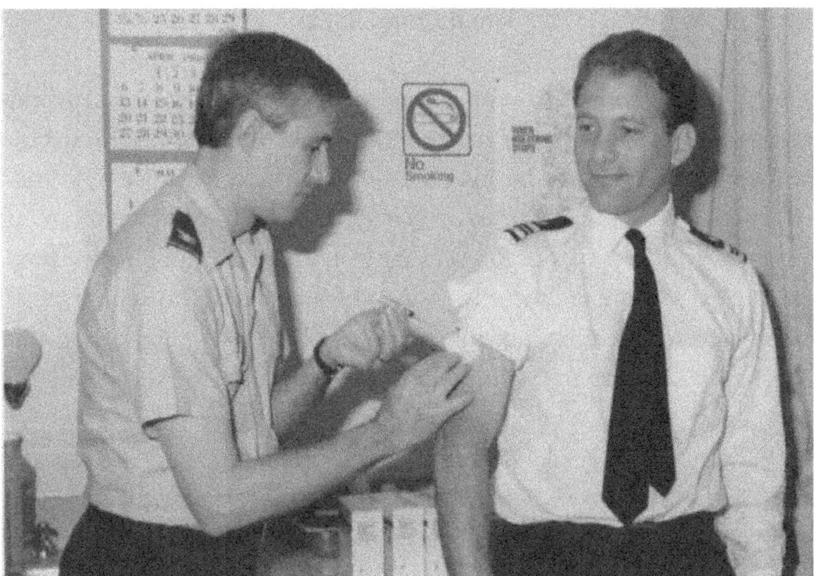

Figure 14.1. Dr. Stephen Hoffman volunteering for a vaccination trial using a subunit recombinant DNA vaccine against malaria. Courtesy of Stephen Hoffman.

nothing resoundingly protective has been developed in almost a hundred years of research. Scientists, including Dr. Hoffman, have tried so called "subunit" vaccines, where fragments of the genome of the parasites are injected.[8] Figure 14.1 shows Dr. Hoffman in 1987 being injected voluntarily with such a subunit vaccine against malaria. After being later challenged with the malarial infectious agent, he showed a decrease in the number of parasites. However, he still developed full-blown malaria. Protection was unsuccessful.[9]

In October 2011 the pharmaceutical giant GlaxoSmithKline (GSK) announced work on an improved subunit vaccine candidate—known as RTS,S. In late 2021 the World Health Organization recommended adoption of the GSK vaccine, which can reduce child mortality by 30 percent. Since there was nothing better at the time, a 30 percent reduction was considered "significant" by the WHO.[10] Yet this is a level well short of the 90 percent protection that good vaccines must achieve.

## The Buzzing Room

After an initial enthusiasm for subunit vaccines and his later disappointment with their poor performance, Hoffman became convinced that the problem could be solved by an entirely different approach. He went back to experiments he had started in the 1990s, using injections of sporozoites taken directly from the salivary gland of a malaria-infected mosquito. That's where the parasites wait to be inserted by the femme fatale, before she gets another bellyful of human blood. The point of forming Sanaria in 2002 was so that Hoffman could go full speed ahead with the live sporozoites.

Hoffman developed methods to produce contamination-free mosquitoes, "aseptic" ones, in the medical jargon. They were sanitized bugs, stripped of bacteria and fungi. After producing the clean mosquitoes, he would let them feed on bags of aseptic malarial blood. Once the mosquitoes were sated, the Sanaria team let the mosquitoes act as bioreactors to produce sporozoites for two weeks. He then irradiated the mosquito-parasite combination to destroy the sporozoites' virulence; in other words, he "attenuated" them. Following this, he anesthetized, decapitated, and dissected the mosquitoes to get their salivary glands. The anesthesia was not so much out of empathy as efficiency. Try catching one flying mosquito at a time out of an acrylic cage of a few thousands and pinning it down on a microscope slide. Sedating all of them at once is a lot simpler. Hoffman next extracted the sporozoites from the glands. He named the isolated live but now attenuated parasites PfSPZ, an acronym that sounds like a bug speaking Gaelic. It stands for "*Plasmodium falciparum* Sporozoites." These PfSPZs were no longer infectious for malaria and would go into humans in sufficient amounts to act as vaccines. At least that was the idea.

It was a wild scheme, first carried out in the original Sanaria labs in Rockville, Maryland, in a small, rented space described in *National Geographic* as "a dismal mini-mall."[11] The operation functioned like an assembly line. In the mosquito room, Hoffman's team would produce the aseptic bugs, feed them, collect them, irradiate, decapitate, and dissect them, extract the parasites, and formulate them into vaccines.

In 2007 I visited the original mosquito room in the dismal mini-mall. Whenever possible I try to visit my clients' labs. Since this was local, I jumped at the occasion. For an old bench-hand like me, there is nothing like the feel and smell of a lab to really get into the science. Sure, I can read about it in documents, but seeing the process at work is invaluable. And, while I left research many years ago, I still love the feel of a bench, a purring instrument, an open lab book, a few pipettes at the ready. It helps me argue my clients' cases in pleadings and hearings.

The first thing I heard when I walked into the mosquito room was the buzzing, the constant buzzing, like that of thousands of planes in bombing formation. The buzzing was everywhere and ever present. The room was outfitted with half a dozen screened boxes where company researchers fed blood to the mosquitoes trapped inside, so that the insects became engorged and red. The meal contained red blood cells infected with *P. falciparum*, the enemy parasite. The mosquito room was kept dimly lit and, other than the high-pitched humming, was quiet. The room was sealed to the outside, and the company's scientists wore protective clothing that gave them the appearance of white ghosts moving from box to box. The careful movement of the attendants in white, walking up to one container and then another, was in stark contrast to the mosquitoes' frenzied gyrations. The insects were feeding on a blood bag placed outside each box, flying off the bag, and then coming back for some more at a different spot. The actions of humans and insects were smoothly coordinated to achieve Hoffman's goal of producing the vaccine.

Hoffman's scientists went about their job with a sense of detachment that was stunning. They seemed unconcerned by the possibility of a mass escape of thousands of mosquitoes containing malaria parasites into the room and onto themselves. Professionals to a fault, they appeared oblivious to the danger. Indeed, members of Hoffman's team had recurring nightmares that something would go wrong and that a bunch of malaria-infected mosquitoes would descend on Rockville and cause havoc in the neighborhood. Thank goodness they have never escaped.

Once the mosquitoes were swollen crimson, they were transferred to the room right next door for two weeks, to allow the sporozoites to develop. Then they were subjected to radiation to attenuate the parasites inside. The irradiation room was barely larger than a closet and held what looked like a larger version of a dentist's X-ray machine. The radiation did not distinguish between the mosquitoes and their parasites; both got zapped equally.

The mosquitoes, however, were disposable. They were but test tubes for the parasites and, shortly after the dosing, were sacrificed and dissected one by one. There was a line of technicians sitting in front of biosafety hoods, placing each blood-filled, aseptic, and zapped mosquito femme under a microscope, and cutting off her head. An experienced technician could decapitate and dissect mosquitoes at a rate of about five to six bugs per minute. I could almost hear Lewis Carroll's Queen of Hearts enjoying herself and commanding: "Off with their heads!"[12]

At the finish of the assembly line, the mosquito detritus, by now a random collection of broken legs and wings, was trashed. In the end, I could no longer hear any buzzing. The dissection room was eerily quiet.

At the time when Hoffman first proposed this idea, he was the most highly cited malaria author in the world. But the other top malaria gurus were skeptical. No one believed that his approach would work. It wasn't so much the science as the logistics. How are we going to make enough vaccine, dissecting one mosquito at a time, to inoculate millions of people? How are live sporozoites going to travel to sub-Saharan Africa, and be stored there waiting for the day of injection? He had trouble getting funding for clinical trials.

But Hoffman is never one to take no for an answer. After all, he was a captain in the navy, survived a plane crash in Kenya, and gave himself malaria in one of several instances of self-experimentation. In this practice he followed a long tradition among curious doctors, such as the South-African Max Theiler, who in 1937 self-inoculated with a yellow fever vaccine, or the Australian Barry Marshall, who in 1984 self-administered the bacterium *H. pylori* to show that it was the cause of gastric ulcers. There are dozens of such examples in many fields of medicine. Hoffman is as close to an undaunted medical hero as I've had the privilege to meet.

Much to everyone's disappointment, Hoffman's first clinical trial showed no significant protection. Because of that, the Bill and Melinda Gates Foundation, an early financial supporter, terminated his funding. Ever undeterred, Hoffman's team figured out why the trial had failed: the sporozoite injections had to be intravenous, not into the muscle. Injecting live parasites into muscles was like dropping parachutists into a dense forest, hoping that they would find the main road. They would get lost. You had to drop them directly on the road, that is, into a vein leading to the liver.

The story goes that with that new insight, he went to Dr. Anthony Fauci, director of NIH's National Institute of Allergy and Infectious Diseases and asked for funding. Fauci is quoted to have said, "Well, it isn't as if you've got a lot of other candidates out there that really, really look good."[13] With Fauci's intervention, a second clinical trial was approved, this one using intravenous injections.

The results were so successful that the stunned skeptics had to concede they had been wrong. None, zero, of the volunteers who had received the attenuated sporozoite vaccine developed the disease after purposefully being infected with malaria.[14] More recent studies in Germany and Tanzania also showed 100 percent effectiveness when the sporozoites were administered intravenously at specific dosages.[15] Not until Hoffman has there ever been a malaria vaccine that is 100 percent effective.

The world noticed. In 2015 Sanaria received close to $50 million dollars from the government of Equatorial Guinea, Marathon Oil, and other oil companies, to sponsor more clinical trials. Prospecting for oil in the tropics is a dangerous business, and everyone, including governments and companies, wants to get rid of malaria. In 2020 Sanaria received €13 million from the European Union Malaria Fund to develop two vaccines for malaria. There is now an international consortium of twenty-seven countries, all focused on malaria vaccines made of attenuated live sporozoites, Hoffman's babies. The Bill and Melinda Gates Foundation came back in late 2023 and is again supporting Sanaria's effort, this time funding an international repository of sporozoite vaccines for more clinical trials.

## Aseptic Mosquitoes

In the early 2000s we started filing Sanaria patent applications on pretty much everything. At latest count, we got them eighteen issued patents in the United States alone. Because of the *Chakrabarty* decision, we patented the live aseptic mosquitoes, cultures, and vaccines containing live attenuated sporozoites, methods of immunization containing these, and methods of producing all these materials, including production of the sporozoites without mosquitoes. We also got them patents on cold storing the attenuated sporozoites so they can survive the trip to malarial regions and are alive and well to do their job when called to do so.

The patent on live "aseptic mosquitoes" might remind you of Pasteur's 1873 patent on "Yeast, free from organic germs of disease, as an article of manufacture," which we discussed in chapter 9. The Patent Office gave Pasteur what he asked for—that is, a claim to a live yeast—although it was later critiqued by scholars as not being the type of invention that could be patented. Well, 140 years later the Patent Office gave Hoffman a patent claim with language that was uncannily like Pasteur's. It was also on a live being "free from germs." Hoffman's patent claim said, "Aseptic Anopheles mosquitoes infected with . . . aseptic sporozoite-stage Plasmodium-species parasites."[16]

Of course, the 1980 *Chakrabarty* decision reassures me that the aliveness of the Sanaria mosquitoes is not an impediment to their being patentable. The next question we need to ask is, Are aseptic mosquitoes carrying aseptic parasites too close to natural materials? Remember *Myriad Genetics*: you can't patent natural materials or anything too close to them. Mosquitoes infected with parasites sure are natural. They fly around the tropics all day long. The main difference between the wild mosquitoes and Hoffman's is that his are "aseptic" and that the parasites inside are also "aseptic." So, is the aseptic quality of the patented mosquito-parasite combination enough to distinguish it legally from the bugs buzzing around at night in sub-Saharan Africa? I am sure of it; and I am not only saying this because I am Hoffman's patent attorney. An aseptic mosquito carrying aseptic sporozoites is useful for a whole bunch of new pharmaceutical things, generating vaccines, for example. Try doing that with a malaria-carrying mosquito hovering around in the rainforests of Equatorial Guinea. You can't.

Plus, the patent on these super-clean mosquitoes was granted on August 12, 2014. That was exactly in the middle of the period between March and December of that same year, when the Patent Office was making up its governmental mind on whether purifying an unmodified natural material was or was not prohibited by the Supreme Court decision in *Myriad Genetics*, the one that dealt with the non-patentability of isolated genes.

As we saw in chapter 11, when I told you the story of purifying Taxol from yew trees, the Patent Office had first interpreted the decision in *Myriad Genetics* as prohibiting any sort of patent on unmodified natural materials. Its initial interpretation was quickly criticized, including by me and my colleagues, as too rigid. So, later the same year, the Patent Office changed its mind. Patent examiners' subsequent interpretation made it easier to get patents on natural materials, like mosquitoes, even without genetically or chemically modifying them. The Sanaria patent on aseptic mosquitoes granted in August of 2014 suggests that the Patent Office had already decided in Sanaria's favor, even a few months ahead of officially announcing its change of heart. For all these reasons, I am very confident that aseptic mosquitoes are indeed human-made "articles of manufacture" capable of being patented.

## Protecting Guillotines

Eventually, scientists at the Rockefeller University and at the NIH bought into the idea of vaccines from attenuated live sporozoites and, together with members of Sanaria's team, are now co-inventors in some of the project's patents. By the way, Sanaria's team includes Stephen Hoffman as well as his wife, Dr. B. Kim Lee Sim, a brilliant, Malaysian-born molecular biologist with close to 150 scientific papers and multiple patents to her name.

The increased collaboration between researchers from Sanaria's team and researchers from other places has also led to a few jointly

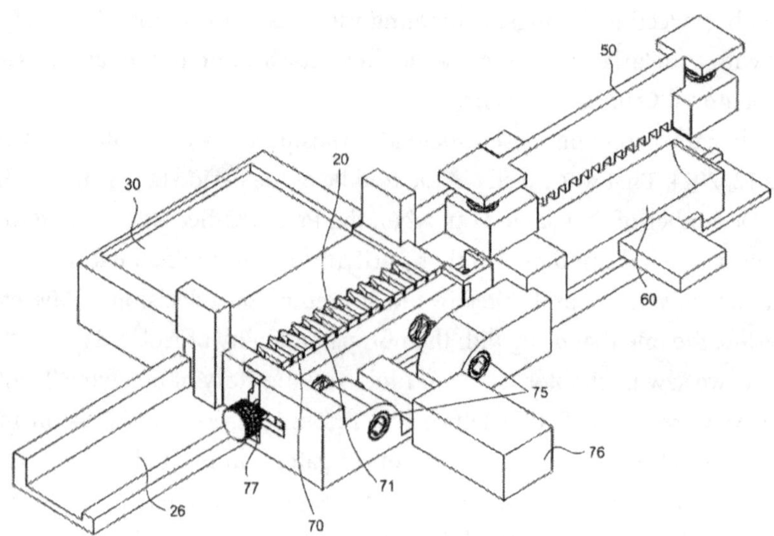


Figure 14.2. The guillotine used to automatically decapitate mosquitoes to extract their salivary glands. Reproduced from Russell H. Taylor, et al. Mosquito Salivary Gland Extraction Device and Methods of Use, U.S. Patent 10,781,419 (filed June 13, 2017) (issued September 22, 2020), fig. 11.

owned patents. One of these is to a mosquito guillotine. By 2020 Sanaria, together with Johns Hopkins University, developed an automatic mosquito-decapitating machine. The elaborate device can cut off heads and then squeeze out the salivary glands of a dozen or more mosquitoes at a time. Multiples of these little contraptions working side by side will streamline the assembly line so that it will look more like a Lilliputian Ford Motor plant than a lab, with robots doing most of the repetitive work. They will run 24/7 with minimal supervision. The robotic line will hopefully solve the early concerns with Hoffman's PfSPZ invention: that no matter how clever the vaccine, it would be impossible to make enough of it by dissecting one mosquito at a time.

I never thought that we would get patents for a mosquito guillotine, but we did. I figure that even the Patent Office, urged by law to be skeptical of everything, was impressed, and issued it.[17] Figure 14.2 shows one of the patent drawings of the guillotine.

Follow the drawing numbers in figure 14.2, and you'll learn how this gizmo works. You drop a group of sedated mosquitoes on platform 30, on the left. Like tiny buses at a Greyhound terminal, they line up, side by side, at the dozen or so bays, 20, in the center. The chopping knife, 50, moves over from the right, comes down on them, and their lopped heads drop on to tray 26, on the left. Simple.

Also in 2020 the Patent Office gave Dr. Hoffman and Sanaria the "Patents for Humanity Award." This is a tribute for visionary inventors who have pioneered ways to provide creative solutions for the less fortunate of the world. It couldn't have gone to anyone more deserving.

I always tell Dr. Hoffman that when (not if) his vaccine succeeds, he will surely win the Nobel Prize. In response, he smiles the same boyish smile I saw that day in my office when he showed up wearing his whites and charmed secretaries and patent examiners alike.

You may ask whether patenting vaccines is a good idea. You may worry about granting exclusive rights to critically needed vaccines during a global emergency like a pandemic. And you may question whether, even if there is no overwhelming emergency but a long-existing public health danger such as malaria, society should allow companies like Sanaria to get patents on lifesaving vaccines. These topics are sufficiently complicated that they deserve a separate discussion altogether, one to which we'll get to in chapter 20.

Meanwhile, let's talk about patenting genetically engineered plants.

# 15

# Plants

Sometime in the mid-1980s I received a call from Pioneer Hi-Bred International, in Des Moines, Iowa. The company wanted me to represent it in a priority of invention fight over who would have patent rights to genetically engineered plants reprogrammed to resist viruses. It was an interference, just like the Enbrel or CRISPR interferences we saw in chapter 4, where two or more parties fight to prove who among them invented first.

At the time, Pioneer was perhaps the best-known seed company in the United States. Because of its prominence, the case had it all: advanced plant genetics; ultimately not one, but two, interferences; an appeal to the Court of Appeals; a close case of "obviousness" of one virus compared to another; subtle patent law concepts; big scientific and judicial names; and major corporate acquisitions. Let's start with a bit of science, this time plant genetics.

## Taking over the Government

You may be surprised to learn that injured plants sometimes develop tumors in their roots and branches. These tumors are known as "galls." The culprit is a bacterium called *Agrobacterium tumefaciens*. This bug is vaguely like the resident bark fungi we met in chapter 11, which fight invading microbial foes and let yew trees live for a long time. Both have evolved relationships with their plant hosts. However, unlike the symbiotic yew fungi, *Agrobacterium* is not just defending the plant and its own niche from invading pests. *Agrobacterium*'s role is much more colonial, predatory almost. It senses a plant bruised by hail or wind, invades,

Figure 15.1 Galls growing on the branches of a rose bush. Source: Wikimedia Commons. Author Cherubino. Licensed under the Creative Commons Attribution-Share Alike 3.0 Unported license.

creates a home for itself, hijacks the prey's machinery to its advantage (something the yew fungi don't do), and comfortably fends off invaders. Figure 15.1 shows several galls, which, courtesy of *Agrobacterium*, are growing on the branches of a rosebush.

It turns out that the invading *Agrobacterium* contains a large circular piece of DNA, a "plasmid" in the jargon, pieces of which the bacterium can easily transfer into plant cells at the site of an injury. This circular plasmid contains many microbial *Agrobacterium* genes. When the microbe introduces part of its plasmid into the plant cell, the microbial genes come along. This is where colonization starts. Subtly, almost unnoticed, *Agrobacterium* puts its own trusted genes into crucial positions of the plant's government; that is, into the plant's genome.

Pretty soon, the bacterial genes that have integrated into the plant genome start meddling with the host's machinery: *Agrobacterium* co-opts the plant's genomic government to its own uses. The disruption mucks up the plant's natural hormonal balance. This intrusion, as is the case with a local government controlled by a colonial power, leads to events that the plant can no longer manage. The invading bacteria, by messing with the hormonal stability of the plant, cause the gall tumors. The tumors weaken the host plant, undermining its lifecycle and facilitating the takeover.

In 1983 Mary-Dell Chilton and her team at Washington University St. Louis (Wash. U. St. Louis) demonstrated that the gall tumor–producing genes could be removed from the microbial plasmid without adversely affecting the ability of *Agrobacterium* to insert other DNA into the plant genome.[1] She co-opted the colonizers to do her bidding. Chilton could now insert any genes into plants, without inserting the tumor-producing genes, and the process was under her regal control. Because of that, she is sometimes called "the queen of *Agrobacterium*."[2]

When the genetics of colonization became clear, we humans, as is our unstoppable wont, took advantage of what evolution had given us, and went further. Once scientists learned how to transfer nutrient genes without causing tumors, they started using the re-engineered *Agrobacterium* plasmid to transfer other, so called "foreign" genes, into plant cells. Researchers started creating all sorts of foreign plant attributes, the variety of which was limited only by their creative imagination. Using *Agrobacterium*, genes were transferred that once inside plants conferred resistance to such harmful things as viruses, insects, drought, or salinity. Other foreign genes allowed plants such as rice to produce unnaturally high levels of vitamins, making them much more nutritious to consumers with vitamin deficiencies. A short roster of plants that have been genetically engineered through the disarmed *Agrobacterium* plasmid now includes soybeans, cotton, maize, sugar beets, alfalfa, wheat, and rice. Long-shelf-life tomatoes, non-browning apples, and purple roses add to the growing list.

The well-known name for such reprogrammed plants is Genetically Modified Organisms (GMOs). The scientific term for modifying plant cells with foreign genes to make GMOs is "transformation." And transformation is precisely what happened to agriculture when, in 1983, *Agrobacterium* started being used to create GMOs. As plant cells became increasingly transformed, so did agriculture.

The transformation of agriculture initially brought along an assorted collection of multinationals, venture capitalists, merger and acquisition specialists, and, of course, patent attorneys. Not far behind that first group came the legislators, lobbyists, regulators, detractors, protesters, and vandals. Those will have to wait until chapter 21, when I will tell you more about patented GMOs. In the meantime, let's talk about intellectual property issues that emerged from the engineering of foreign traits into plants.

## Turf Battles

The patent hurdles with plant genetic engineering were not about eligibility. Since the *Chakrabarty* decision of 1980 and its progeny allowed patents on human-modified living things, genetically engineered plants became eligible "articles of manufacture." Once the basics of modifying plant cells with *Agrobacterium* were mastered, and scientists could introduce new genes without causing gall tumors, the issues were which plants, which genes, which traits? There was an explosion of new applications using the new *Agrobacterium* platform for plant genetic engineering.

The main problem in this area was how crowded it became with commercial competitors. The labs of academia and companies made fundamental discoveries, and, after that, the behemoths of the industry started applying to the USDA for approval to launch GMO products. Everyone saw green gold in those cultivated fields, packed, row after geometric row, with engineered corn, soybeans, or rice. Everyone wanted at least a few of those rows. The products, once approved, would be bags of genetically engineered seeds that had been finely tuned for this or that trait. Just who had the IP rights to pack these seed bags and sell them to farmers became the turf battles of the age.

Quickly, way before the first sales, everyone started fighting about patent rights. And many such fights were interferences. The questions in plant genetics were several: Who was first with methods of introducing foreign genes into plant cells? Who was the first to do so using the *Agrobacterium* plasmid? Who was first in introducing viral resistance, drought resistance, or insect resistance? Dealing with these questions was a boon for biotech patent attorneys. The science was complicated, the area of law was new, and most judges, while not very knowledgeable about plant genetics, were curious and open minded. Even better, the supply of patent attorneys who had the necessary qualifications was low, and the demand was high. Many potential agricultural clients came knocking during those years: Pioneer Hi-Bred, Archer-Daniels-Midland, and Bayer Crop Science, to name but a few.

Pioneer had been founded in 1926 by Henry Wallace, the future vice-president of the United States. At the time of its birth, Pioneer sought to take advantage of then recent discoveries in the science of hybrid seeds. The creation of hybrid seeds was a breakthrough in plant breeding that had nothing to do with modern genetic engineering. It happened eighty years ahead of the re-engineering of *Agrobacterium*'s circular plasmid by Mary Dell-Chilton, which led to the exquisite precision of our modern era of plant genetics. Hybrid seeds were made by good old-fashioned plant breeders out in the sunny cornfields of Iowa.

What are hybrid seeds? Well, when two distinct populations of highly inbred corn plants—one male, one female—are crossed with each other, the result is an "F1," also known as a "First child," or "Filial one," generation plant with so-called hybrid vigor, or heterosis.[3] These F1 plants show improvements in height, growth, and yield that their individual parents don't have. The beauty for plant breeding companies such as Pioneer is that the second, F2, generation seed derived from the F1 hybrid does not grow "true to form." That is, the second generation does not have the same vigor as its hybrid parent; many revert to the less vigorous grandparents. (Any grandparents reading this know exactly what I mean.) Thus, Pioneer's customer farmers could not generate vigorous hybrid seed after their planting season; only Pioneer, with its carefully guarded inbred parental lines, could. The farmers would therefore need to go back to

Pioneer every season for more F1 seed. Henry Wallace, the company's founder, recognized that the hybrid seed developments of the early twentieth century would capture returning customers for years to come. As Ira Gershwin said, "Nice work if you can get it." And it was nice work for Pioneer Hi-Bred until the advent of modern plant genetic engineering in 1983.

Around that time, a younger generation of Pioneer plant breeders was starting to learn about contemporary biotechnology. They realized that one could insert a trait—say, pest resistance—into an inbred parent. The engineered parent would not only be resistant to the pest, but its seed could be saved from one planting season to the next, and then replanted. The customers would not have to come back to Pioneer every year for more F1 hybrid seeds, threatening a fundamental business model of the company. And the *Chakrabarty* decision from the Supreme Court had just announced in 1980 that you could patent living things, something unimagined by the old guard in Des Moines. If a competitor got a patent on corn plants engineered to be resistant to some nasty plant virus, Pioneer could not sell that without a license. The times they were "a-changing," as Bob Dylan would say.

So, Pioneer decided that it needed to get a modern biotech patent lawyer, and some patents of its own. When a competitor holds patents that you might need, it is good practice to have a few in your hands that it might need. Then you can start horse-trading. And that's why, in the late 1980s, I received a phone call from Pioneer.

"Could you represent us?" asked its general counsel.

"Who are your potential adversaries?" I asked back.

"Monsanto," was his answer.

"Last I checked, we don't represent them, but you know how quickly things move," I said. "I'll call you back tomorrow to confirm."

It turned out that Monsanto was not a client and had not recently bought one of our clients, so we could represent Pioneer against the former. Monsanto was, as I call such powerful adversaries, an "anti-client." I have learned that sometimes it is better to have one or two litigious anti-clients than ten small clients. It's a full employment act for the law firm.

At the time, Monsanto was emerging as the big gorilla in the plant biotechnology field. It had been around since 1901 in St. Louis, Missouri, and for many decades had a profitable life as a chemical and pharmaceutical company. In the mid-1970s Monsanto developed the herbicide Roundup® for non-selective weed control. This lack of selectivity meant that while the herbicide killed weeds, it also killed some of the main crops it was supposed to protect. This was not a desirable thing. Sales of Roundup started weakening due to concerns about selectivity and about health risks. The herbicide itself went off patent in 2000 and became a generic, not a good thing for Monsanto. Others could produce it and the price would quickly drop.

As the marketing problems with Roundup mounted and the loss of its patent exclusivity became imminent, in 1991 Monsanto took licenses from Agracetus, a California company whose scientists had invented Roundup-Ready® soybeans. These are genetically engineered soy plants that are resistant to Monsanto's herbicide; thus, in one fell swoop, Roundup became highly selective. Roundup-Ready soybeans would henceforth grow happily in fields sprayed with Roundup, while their neighboring weeds died. Sales of Roundup started rebounding, and sales of Roundup-Ready crops took off all over the world. Because of these maneuvers, Monsanto became a very successful agricultural biotech company. And because of the many lawsuits that the maneuvers engendered—whether with farmers, consumers, or competitors—Monsanto also became a much feared and maligned multinational company. It was a perfect anti-client.

## Fights for Alfalfa

Monsanto became my adversary in the late '80s, in the early stages of its approach to the biotech world. By then the company had established a productive research collaboration with Professor Roger Beachy at Washington University in St. Louis. Beachy was an expert in plant biotechnology and a specialist on how to make plants resistant to viral diseases. At some point in the future, Monsanto wanted to sell virus-resistant plant seeds, so the company was investing in basic research. Beachy was their man.

By the time I received the phone call from Pioneer, both my new client and Monsanto had, less than six months apart from each other, filed separate patent applications on creating virus-resistant plants by genetic engineering. The main inventor on the Pioneer side was Loretta Sue Loesch-Fries. She had been a scientist at Agrigenetics, a small Wisconsin company, where she did the basic research involved in her patent application. While she was there, Agrigenetics and all its IP were bought by Lubrizol, an Ohio company suddenly interested in making specialty oils by genetic engineering of plants. But that did not last long. When Pioneer called me, it had recently bought Loesch-Fries's invention from Lubrizol, including her notebooks, data, and biological materials. She thus became a scientist at Pioneer. Perhaps dizzy from the constant and involuntary changes of employer, she sought stability and became an adjunct professor of virology at the University of Wisconsin.

The main inventor on the Monsanto side was Roger Beachy at Washington University of St. Louis. The two professors, Beachy and Loesch-Fries, knew each other and had collaborated on common scientific interests. Ironically, in 1990, a few years after their separate patent filings, they even published a paper together on the very field of what would shortly become their hotly contested battle: antiviral resistance in plants.[4] They were respected scientific colleagues from top universities but, due to different corporate ties, became formal opponents during the five years of the litigation. Their corporate funders had put them in an awkward legal situation.

I never met either personally, although from what I know of them, I would have liked to, and not just because they were smart and creative scientists. Beachy was an opponent, so I could have only met him had I taken his deposition, but that never happened. Loesch-Fries was a client, but the many lawyers between her and me made it hard to communicate directly. The first line of lawyers was at the University of Wisconsin, and they were both inside and outside counsel. Then came the lawyers for Pioneer, her corporate sponsor, and then came I. To get through to her I needed to alert layers of legal talent before even picking up the phone. So, beyond her genetics work, I didn't get to know her well.

In June 1991 the Patent Office declared an interference between Loesch-Fries and Beachy, and the race was on to figure out who had first invented antiviral plant protection. The battlefield was populated with plant viruses. The most common of these are the so-called mosaic viruses, a name they acquired because of the mottled look their plant victims take on after becoming diseased.

Figure 15.2 shows side-by-side photographs of a tobacco leaf on the left and alfalfa leaves on the right, infected with mosaic viruses.

There are thousands of mosaic viruses, and their variety is as diverse as the plants they infect. Here's a (very) short list of plants targeted by the mosaic viruses; the list is organized poetically rather than taxonomically: peas, papaya, potato, passion fruit, peanut, poinsettias, peaches, parsnips, and peppers. Farmers know that viral infestation of such crops leads to costly epidemics, which diminish yields, quality, and food supplies.[5] Anyone who could figure out how to prevent mosaic viruses from infecting plants, especially commercial plants like corn or alfalfa, would hit the jackpot. Beachy and Loesch-Fries—or, more accurately, Monsanto and Pioneer—were competing for the mother of all plant antiviral patents.

When I studied Beachy's and Loesch-Fries's patent applications, I was immediately struck by the widely different scope of experiments the two teams had included in their filings. Beachy had examples of plant resistance to half a dozen different viruses, such as those that infect tobacco, alfalfa, potato, tomato, cucumber, and cauliflower. Loesch-Fries also had several examples, but they were all for one virus, the alfalfa mosaic virus (AlMV). After getting through multiple layers of lawyers, I chatted with Loesch-Fries. She quickly educated me on some very basic differences between the alfalfa virus and all other viruses. AlMV stands out because it is a more complicated virus than the rest. How complicated? The answer for me was, complicated enough to be in a separate priority contest altogether. If we could prove that AlMV was "non-obvious" over plant viruses in general, we could convince the Patent Office to set up a second

Figure 15.2. Leaves infected with mosaic viruses. *Left side,* tobacco leaf infected with tobacco mosaic virus, TMV. Source: USDA Forest Service. Author R.J. Reynolds Tobacco Co. Slide Set. Public domain; *right side,* potato leaves infected with alfalfa mosaic virus, AlMV. Source and author: Howard F. Schwartz, Colorado State University, bugwood.org. Licensed under CC Attribution 3.0 Unported license.

interference. And it seemed that we could readily win that one on the dates of invention.

"Could we argue that AlMV is 'non-obvious' over all the other viruses that Beachy describes?" I asked Loesch-Fries.

"Yes," she replied. "Let me send you a paper that Roger himself wrote about how different they are. He uses tobacco mosaic virus [TMV] as representative of all others."

Sure enough, none other than Roger Beachy, almost two years after he had filed his patent application on antiviral protection, had published a paper in which, to our legal delight, he had presented a study comparing alfalfa virus to tobacco virus. He explained (the emphases are mine): "The objective of the current study was to determine if the phenomenon of genetically engineered cross-protection could be extended to a different group of plant viruses. We have chosen alfalfa mosaic virus (AlMV) for this study. *AlMV differs from TMV in many respects. . . . TMV and AlMV are clearly distinguished by their morphology, genome structure, . . . and modes of transmission.*"

And, after he described the thorough study that his team had done comparing the two types of viruses, Beachy concluded his paper by saying: "We show here that genetically engineered cross-protection can be demonstrated *with a virus [AlMV] which is clearly distinguishable from TMV* in its morphology, genome structure, gene expression and early steps in replication, and that it can be achieved in different host plants for the virus."[6]

Beachy kept emphasizing how AlMV—the alfalfa mosaic virus—was "clearly distinguishable," belonged to a "different group," and diverged from TMV in "many respects." He also questioned whether cross-protection "could be extended" to AlMV, implying that such a result was far from certain.

Beachy had said nothing of the kind in his patent application of 1985. He didn't have to, really. He had achieved antiviral protection against a half dozen viruses, and he didn't need to focus on AlMV. He ignored any differences. If you read his patent descriptions, AlMV was just one more, lost in the handful of viruses he exemplified. Until Loesch-Fries explained it to me later, I couldn't tell that to a plant molecular biologist, AlMV really stuck out like Michael Jordan in a peewee league. Beachy did focus on the differences between the viruses two years later, in his 1987 paper. His belated admission was a blessing to our clients. Beachy had unwittingly made our case.

With his publication in hand, we proposed to the Patent Office that there should be two simultaneous interferences running in parallel. The first one would be on the broadest possible definition of a plant virus; it would include protection against all viruses. In the jargon, the first interference would be on the "genus" of all viruses. The second interference would be specifically on protection against the "species" of AlMV only. You may remember from Chapter 12 that I there illustrated the concept of a patent "genus" as a collection of all chairs, of which dining room chairs were a "species." Same here with respect to the genus of all viruses, with AlMV being a species. Using Beachy's own words, we argued that AlMV was a "significantly different" virus than, and one that was "clearly distinguishable from," the genus of all viruses. Because of that, AlMV was a different invention. We added that even our esteemed opponent, Professor Roger Beachy of Wash. U. St. Louis, admitted as much in 1987. This is conclusive evidence, we said, that two years after filing his own patent application he still believed it. So, we concluded, AlMV is non-obvious over the rest of the plant viruses. It's a separate invention and should be in an interference of its own.

You could have heard Monsanto's complaints about this strategy all the way from Missouri to DC. It was nonsense, the lawyers said. All viruses are similar; the genetic engineering is similar; both teams used disarmed *Agrobacterium*. A second interference would be a waste of time and money. And, they added, Beachy ultimately showed in that very 1987 paper that one could achieve plant protection against AlMV just as nicely as against TMV. So, they said, the early concerns that Pioneer is now echoing have turned out to be unfounded.

But the Patent Office Judge ruled in our favor. It doesn't matter, said the judge, that ultimately the antiviral resistance worked with AlMV just as it did with TMV. That's all hindsight. At the time of the filing dates of the patent applications, two years before those experiments were completed, everyone believed that AlMV was significantly different. It is the expectation at the time of filing that matters, not experiments carried out years later. At the time of filing, there was no expectation that things would work well with AlMV. And such lack of expectation is a legal tell-tale sign of non-obviousness. Thus, antiviral protection against AlMV is non-obvious compared to antiviral protection against TMV, and it's a separate invention.

On December 15, 1992, the judge set up the second interference. Less than two weeks later, on December 30, 1992, Agracetus bought the pending patent application of Loesch-Fries from Pioneer and became my client. You may recall that Agracetus was the company that invented Roundup-Ready soybeans, a genetically engineered crop that made Monsanto's Roundup highly selective and again profitable. In 1991 Monsanto had taken a license to the Roundup-Ready soybeans and no doubt had taken a liking to Agracetus. I wasn't too surprised then that Agracetus bought Pioneer's pending application. It must have thought that there would be something else that Monsanto might want from it in the future. I can only surmise that Agracetus was watching in the wings and, once it saw that there would be a second interference focused only on alfalfa, jumped and made its acquisition. Events would soon show that this was a wise move.

Beachy's dates of invention for the first interference, the broadest one that included all viruses, were earlier than Loesch-Fries's. He had invented first, and she could not win the first interference. In contrast, her dates of invention for the second interference, the narrower one on alfalfa, were earlier than Beachy's, so Monsanto could not win that one. With the second interference pretty much assured, we conceded the first one. Monsanto won it, and the company would soon receive a patent for antiviral protection against all viruses. Yet Monsanto was now facing the real possibility that Agracetus would win the second interference, on antiviral protection against AlMVs. If that happened, then Monsanto could not sell antiviral protection against AlMV infection without a license from Agracetus. And Agracetus could not sell antiviral protection against AlMV infection either, without a license to the broader patent of Monsanto. This conundrum is known in patent law as a case of "blocking patents." Let me explain.

Let's assume that I invent TV and I get a broad patent for "television." A few years later, after much research and creativity, you invent color TV. The Patent Office decides that color TV is non-obvious compared to plain old TV, and it gives you a well-deserved narrower patent on "color television." The patent on TV is on the genus, and the patent on color TV is on one of the species. Our patents are now blocking one another. Neither you nor I can sell color televisions. My patent on all televisions lets me keep you and the rest of the United States out of the television market. But your patent on color television does not let me sell a single unit of color TV either. For both of us to sell color TVs, we need a license from each other. To unblock our patents, we can negotiate a so-called cross-license. Only then can the two of us sell color TVs.

If our new client Agracetus succeeded in winning the second interference, the two companies would be blocked when it came to commercializing antiviral protection against AlMV. Nobody could do that without a license from the other. That was our plan, and we were almost there. But Monsanto was not ready to give up so easily. It appealed the decision that

AlMV was non-obvious compared to other viruses to the Court of Appeals for the Federal Circuit (CAFC).

The hearing at the court has stayed with me. It is one of the things that I remember most vividly about this fascinating case.

## Judge Rich

My legal team and I showed up at the beautiful courthouse on Lafayette Square at about nine o'clock in the morning. This was the same court where I had won the *Wands* case in 1988, the one on the enablement requirement, which I told you about in chapter 12. I was not nearly as nervous this time.

We had plenty of time to complete all formalities. Once inside the entry hall, we rushed to see the order of the cases and who would be our three judges that morning. The custom at the CAFC is not to let parties know ahead of time what judges will sit for what hearing. You find that out the morning you arrive. That way, you are not tempted to write your brief to accommodate the views of any one judge. You have a few minutes to prepare mentally for questions that might come from one judge or another, but that's all. Since there are usually three or four hearings per morning, you also do not know until you arrive if you will be first or last. And, you don't know in what courtroom you'll be arguing until you look at the posted day's agenda.

I could not believe our luck when I walked over to the glass-enclosed wall board and saw that the judicial panel for my case, *Loesch-Fries et al. v. Beachy et al.*, had three of the most experienced and erudite judges of the court: Sheldon Plager, Paul Michel, and Giles Rich. Of these three, Judge Rich, at ninety-two, was by far the most senior. He had been appointed by President Dwight Eisenhower in 1956. He was known for his quick and brilliant insights into complicated cases. Indeed, he was famous for having written the 1979 *Chakrabarty* opinion reversing the Patent Office's view that living beings were not eligible for patents. His opinion in that case was a scholarly treatise of patent law and was closely followed and affirmed by the Supreme Court in its own *Chakrabarty* decision a year later. I had never argued a case before Judge Rich and was delighted to finally be able to do so. It was probably one of the last cases he heard, as he died three years later.

I went first, loaded for bear. I was ready for a lengthy debate with the judges as to why my client Loesch-Fries was entitled to a separate interference on antiviral protection against AlMV. Why alfalfa mosaic was non-obvious over, say, tobacco mosaic, and because non-obvious, it was a distinct invention and worthy of a separate priority determination.

I started: "Good morning and may it please the court. My name is Jorge Goldstein, and I am counsel for Party Loesch-Fries," I said. "I will take fifteen minutes and reserve five for rebuttal." The judges nodded.

"It's our position that AlMV is a very different virus than all other viruses in this case . . ." But I didn't get another word in.

Judge Rich interrupted and, lifting a copy of Beachy's paper of 1987, said with a mischievous smile, "Your opponent says as much in his own paper, doesn't he, counselor?"

I was stunned and delighted. "Yes, Your Honor," I responded. "Even two years after their own filings, our opponents admitted that AlMV is distinct." I looked back at our opponents' table and noted a rising paleness in the faces of Monsanto's lawyers.

"Is there anything else you wish to add, counselor?" asked Judge Rich, inviting me to just end my efforts right there and then. In his old age, Judge Rich had no time to waste with details. He had cut right to the chase. At that instant I felt that I had won the case.

"No, Your Honor," I said. "I think you understand our position quite well. We are entitled to a separate interference on AlMV. This court should therefore affirm the decision of the Patent Office. Thank you." Then I added, "Oh . . . and I may use my five minutes of rebuttal, if I need them."

And I sat down. I had learned over the years that if there is nothing to say, I shouldn't say anything. Judge Rich was as sharp and witty as he had been throughout his forty years on the bench. Any concerns I might have had about his mental quickness were swiftly dispelled that morning. His playful grin has stayed with me.

Monsanto's lawyers had very little to say. It was clear which way the wind was blowing that spring morning. A few months later the court handed down its decision. It ruled in Loesch-Fries's and Agracetus's favor. In the last few lines of the opinion, written by Judge Plager, the court said, "[An] article on which Beachy was a co-author suggests that those skilled in the art would not necessarily have expected AlMV to work as a substitute for TMV."[7] Beachy's article had clinched it for the court. As a result, both sides would now have blocking patents.

Because of that, it didn't surprise me to find out that a few months after the oral hearing, and even before the written opinion came out, Monsanto bought Agracetus and all its assets. I guess that the Monsanto lawyers in that courtroom must have called home, and said something like, "We're going to lose this one." As Agracetus must have hoped when it bought the Loesch-Fries pending application, it now had something that Monsanto needed: the freedom to commercialize antiviral protection in plants against AlMV infection. So, Monsanto bought Agracetus, owning the Loesch-Fries patent and its owner. Not only did Monsanto purchase the patent freedom it needed, but Agracetus today is one of Monsanto's central research facilities, in Middleton, Wisconsin.

Pioneer Hi-Bred, my original client in the alfalfa litigation, was bought by DuPont in 1999, three years after the litigation ended. It became DuPont Pioneer. Pioneer has fully embraced the concept of patenting plants and seeds, whether genetically engineered with the plasmid of *Agrobacterium*, or hybridized by the old-fashioned techniques of Henry Wallace, the company's founder. It now holds more than 3,000 US patents on inbreds, hybrids, and GMOs. The largest number of these is on their favorite grain, corn. The very first of these patents, 4,731,499, was filed by me in 1987, shortly after the young guard in Des Moines decided that they had to get a patent lawyer who understood genetic engineering.

At this writing, Roger Beachy is a professor emeritus at Wash. U. St. Louis and a chief science officer at Kultevat, a company that uses genetically engineered dandelions as a source for natural rubber. Loretta Sue Loesch-Fries is an associate professor at Purdue University, where she continues her research on the alfalfa mosaic virus. As I was completing this chapter, I talked to her to make sure I remembered things correctly. She was delighted to recall the interference days of long ago. She graciously reminded me of all the companies that had laid claim to her invention: she had made the invention at Agrigenetics, she said; it was sold to Lubrizol, then sold to Pioneer, then to Agracetus, then to Monsanto. Many years later, Monsanto was sold to Bayer, and her patent ended up in the coffers of the German multinational. We were both bewildered by the many mergers and acquisitions. Her invention of antiviral protection for the alfalfa mosaic virus is a classic example of property that, once out of her head, took on a life of its own. It was secured by patent applications, folks fought over it, judges ruled on who was entitled to it, and companies bought it and sold it. It was just like other property, except it was intellectual.

I asked her if she had ever seen her invention commercialized, and she said no. "There are other ways now of providing viral resistance to plants than the one I invented in the 1980s, so my technology has been leapfrogged," she added, not without a touch of sadness.

She and Roger Beachy have remained good colleagues. They have probably forgotten the adversarial positions they took (or at least that their lawyers took) three decades earlier, during the patent interference litigation on alfalfa mosaic virus.

# 16

# Mammals

We end Part III with a tale of patenting animals, big animals. A quote from Lewis Carrol's poem "The Walrus and The Carpenter" seems appropriate:

"The time has come," the Walrus said,
"To talk of many things:
. . . [Like] why the sea is boiling hot—
And whether pigs have wings."[1]

No one has yet genetically engineered pigs to fly, and I hope they never do. But how about genetically engineered cows that produce human hormones in their milk? Yes, scientists have made such cows, and I have helped to patent them. I have obtained patents on plenty of living things: genetically engineered microbes, genetically engineered seeds and plants, aseptic mosquitoes—you name it. If it moves and reproduces, I have probably patented it.

In the process, I have learned that life is not always easy to see. To appreciate the aliveness of yeasts, you must look under the microscope and watch them move around. To see the vitality of a virus-resistant alfalfa seed, you must plant it and wait a few weeks for it to sprout and develop into a full-grown plant. Mosquitoes at least show their liveliness by speedy movements and buzzing noise. Yet the time I filed for patents on a genetically engineered cow takes the prize. Nothing beats seeing a patented calf being born to get the gist of what it means to patent life.

## Of Two Minds

My client was an Argentine biotechnology company, which I will call BuenGen. Its labs were in the middle of a crowded residential neighborhood of Buenos Aires (BA), a few miles from the business center. Behind a modest façade fronting an old cobblestoned street in the shadow of an elevated freeway was a modern set of laboratories that looked like it had been transplanted from a biotech company near Highway 101 in Silicon Valley. BuenGen considered itself the pride of Latin American biotech. It focused its efforts on countries where a drug that it was manufacturing was not protected by other companies' patents. These were mostly places in the Global South like Vietnam, Cuba, or Ecuador. They were countries that the pharma companies of the Global North had overlooked.

The pharma industry of Europe and the United States makes most of its profits from drugs patented in the big markets of the North. That's where well-to-do consumers will pay top prices for the latest cures. The less-developed regions of the world have smaller purchasing power, so they are less profitable. And, even if the industry gets patents there, the likelihood of getting a court to enforce a patent against a small local manufacturer to let a multinational sell it at a higher price is, let's say, not a top priority of the countries' judicial or political systems.

Those off-the-patent countries were BuenGen's main markets. Argentina is not a country with a strong patent system, so, free of worries, the company manufactured others' pharmaceutical drugs in its labs in that cobblestoned neighborhood. There BuenGen copied and made many blockbuster drugs that were not patented in Argentina. It sold these drugs in places that had been overlooked by the inventing companies, or whose patent systems were toothless. It paid no license fees to anyone, not at home, not anywhere. It called itself a "generic biotech company." Companies in the United States who knew of BuenGen, and whose drugs it copied, however, had other, let's say, less flattering names.

BuenGen was not a true generic pharma company, like, say, Israel's Teva Pharmaceuticals. Teva copies drugs patented by others, but before it can sell a dollar's worth, it must litigate against the others' patents and try to invalidate them. Such lawsuits take lots of money and years of effort. Sometimes the litigation succeeds; sometimes it doesn't. If it fails, Teva must wait until the patents expire, which may be years in the future.

While BuenGen was brazenly copying, it was not doing anything illegal. I would call its strategists legal copyists. They were masters at avoiding the world's patent maze. So, I was surprised and confused when they hired me to get them patents. Had they seen the light? Well, I came to understand that they had not seen the full glow but maybe had caught a glimmer.

Turns out that inside the company there was a group of researchers who was doing some serious biotechnology innovation. Like all creators and inventors, they—surprise, surprise!—did not like others to copy their breakthroughs. The company was of two minds when it came to patents. It had a corporate identity with a split personality, one that I have seen often and not just in the developing world. I can best sum it up as: "patents for me, but not for others." I learned years later that the internal debates on whether BuenGen should obtain patents or ignore them would go on for hours. In the end, the "no patents" faction won out, and I was eventually relieved of my legal responsibilities, or, in the vernacular, I was fired. But not before I did a decade of interesting work.

I got a call in 2003 asking me if it was possible to patent cows. The chief of research told me of a calf that a team of researchers had genetically engineered to produce human growth hormone in its milk. This is the hormone that stimulates growth and development in humans. The transgenic cow was the co-invention of scientists at BuenGen and Dr. Lino Barañao, a member of the National Scientific and Technical Research Council of Argentina (CONICET). The reprogrammed cow allowed the company to separate and purify the hormone from the milk and commercialize it for the treatment of deficiencies, such as short height. It was just like making the hormone in genetically engineered bacteria, except these were genetically engineered mammals, a lot more massive and a lot more complicated. But both were biofactories for the human protein. I knew by then that following the 1980 decision in *Chakrabarty*, patents had been issued on human-made small animals, such as modified oysters. So, I said, "Sure, you can patent cows, as long as they're artificial enough; but you can do that in the US, not in Argentina."

Understood, he replied, BuenGen wanted a US patent. OK, I went on, but remember that a US patent on a cow will only protect your IP in the cow in the United States, not if it's living in Argentina. Patents are national. Only an Argentine patent will protect the cow grazing on the fertile Pampas, and as you know you won't be able to get such a patent down there. Getting animal patents in Argentina is not possible.

BuenGen seemed unconcerned. Maybe the company hoped to do a deal with an American company to produce human growth hormone in the United States using patented cows living in a Wisconsin dairy farm. Maybe all it wanted was the prestige of having a US patent on such an ingenious Argentine invention. Just in case, we made sure that we not only tried to get a patent claim on the cow itself but that we filed for a claim on a method of purifying human growth hormone from the milk of transgenic cows. A method like that patented in the United States would at least allow me to go to a US federal court on behalf of BuenGen and block importation of growth hormone into the United States from any country where such a procedure was used to purify the hormone.

## Non-Obvious Cows

We prepared and filed a patent application with multiple claims. We gave a Julia Child–type recipe for how to make a transgenic cow in six easy steps: (1) get a human growth hormone gene and, together with other specialized DNA sequences, insert it into cow mammary cells growing in a petri dish; (2) remove the nucleus from a cow egg cell; (3) fuse one of the mammary cells carrying the human gene with the egg cell from which you have removed the nucleus—this will result in a one-celled embryo, which carries the human hormone gene in its new nucleus; (4) implant the embryo in the uterus of a surrogate cow; (5) monitor the pregnancy through the birth of the transgenic calf; and (6) let the calf grow to lactation, milk the grown-up cow, and—voilà!—her milk will have human growth hormone in it.

Don't worry if you could not follow any of this. I just wanted to give you a high-level idea of how detailed a biotech patent needs to be written to fully describe and "enable" the product sought to be patented. In fact, the actual patent descriptions were a lot more complicated than what I just described.

One more thing: no, you cannot drink the milk to get its hormonal effects. Your stomach will break down the hormone together with the rest of whatever you've ingested. You must extract the hormone, and you may purify it by the method that we described in the patent. Once pure, you can administer it, preferably by injection.

We filed the application and waited for about two years, the normal length of the examination queues. We were curious to see what would happen. When we received the first examination report, we sighed with relief when we saw that the Patent Office had no problem with patenting transgenic cows. Perfectly all right to get such patents, it said. This cow is just as engineered as the transgenic microbes of Dr. Chakrabarty. It is human made, artificial enough, and eligible for patents. But . . . there was a catch.

A cow that produces human hormone in its milk, it said, is "obvious" compared to a goat that produces the same hormone in its milk. The Patent Office had found a scientific paper showing that such goats had already been made and were quite happily producing human growth hormone. The goats were in the "prior art," as the law calls it, and there was nothing inventive in making one more mammal, albeit a somewhat bigger one, to do the same thing. Nothing patentable in a mere "obvious" development, said the Patent Office. It agreed that such a cow had never been made; that meant that it was "novel." But the cow was not patentable because making such a cow was well "within the skill of the art."

This is a good time to further explain two fundamental concepts of patent law: "novelty" and "non-obviousness," or, as the Europeans call the latter, "inventive step." It is established dogma that you can only get a patent on things that are not already in the public domain. You cannot prevent the public from using what it already had before you got on the scene. If, before the filing date of BuenGen's patent application, a scientific publication or a presentation at a conference by somebody, anybody, anywhere in the world, described a genetically engineered cow that made human growth hormone, the Argentine cow would not be novel. The Patent Office would send us a rejection for "lack of novelty," and we could not get our client its patent, no matter whether we took the case all the way to the Supreme Court. But that is not the end of the inquiry. Obviousness or non-obviousness comes next.

If, let's say, no one in the world had ever made or described a plastic barn door, but there was in the public domain a wooden one, the Patent Office may still reject an application to the plastic door. It would now say that since wooden doors had already been made, making a barn door in any material would also be in the public domain. It would send a rejection saying that while novel, it would still be "obvious" to make a plastic door. Think of obviousness as the legal penumbra hovering over novelty. The idea is that not only can you not get a patent on something that is already in the public domain, but you can't get a patent on anything in the penumbra either.

The difference between obviousness and lack of novelty is that if possible, you can argue your way out of the former but never the latter. To overcome a rejection for obviousness, you might be able to muster evidence that there are some unexpected advantages of a plastic door over a wooden one. The advantages must be unexpected, not something like "The plastic door is lighter and does not require periodic painting." That's not unexpected. Unexpected may be that a plastic door inhibits home robberies far better than a wooden one. If you can bring that type of proof to the Patent Office, you can then argue that plastic doors are "non-obvious" (or "inventive") compared to wooden ones.

That's what happened in our case, with the goat and the cow. The Patent Office said, "A goat? . . . A cow? What's so inventive about doing in a cow that which has already been done in a goat? It's obvious." I had never had to face such a surreal rejection from the Patent Office: that a cow would be obvious compared to a goat! This had nothing to do with genetics. It was more Dalí than Darwin. The two animals being compared had become abstract legal entities. How would we respond to this rejection? What was unexpected about the Argentine cow?

We spent hours brainstorming. We knew that our cow produced more human hormone per liter than the goat in the prior art. Could we maybe argue that this was "non-obvious"? Well, not really, said our client. Cows, being bigger than goats, produce all sorts of things in their milk in larger quantities than goats do. That would not carry the day. Not unexpected.

Then, one smart member of my legal team of scientist-lawyers studied up on both human and bovine growth hormones. We are both mammals, she concluded, and, not surprisingly, our hormones are chemically similar. Human growth hormone will act on a cow almost like the cow's own growth hormone would. So, she said, since the transgenic cow was carrying a higher load of active hormone than the amount it normally had, why wasn't the cow enormous? Why didn't it suffer from what is known as "gigantism"? This is a condition that results in unusually tall humans who suffer from an excess of growth hormone. The most famous example is Robert Wadlow, known as the Giant of Illinois, who, when he died at age twenty-two in 1940, measured almost nine feet. And much to our delight, gigantism is not limited to humans. Blossom, a giant Holstein cow that died in Illinois in 2015 at six feet, four inches, made it into the *Guinness Book of World Records* as the tallest cow ever. There must be something in the water in Illinois.

Anyway, the argument won the day. We convinced the Patent Office examiner that the lack of gigantism was unexpected and that a normal-sized transgenic cow was "non-obvious" compared to its smaller cousin, the prior art goat. We modified the patent claim to include levels of growth hormone in the milk that were even higher than those expected from the size difference between goats and cows. The prior art goats were not gigantic either, but the unnaturally high level of hormone in the cow's milk carried the day. We got a patent for BuenGen shortly after that.[2]

A few years after the cow patent issued, the Court of Appeals decided a case dealing with the patentability of another mammal, Dolly the Sheep.[3] Dolly, you may have read, was a perfect clone of her sister. The researchers had taken cells from the sister's ear and cloned them, developed them into an embryo, transferred the embryo into a surrogate mother, and saw Dolly be born alive and well. She was an exact copy of the sibling whose ear had led to her life. The Scots who invented Dolly filed a patent application. Yet first the Patent Office and then the Court of Appeals refused to grant them a patent. Their main problem, said the court, is that in their very application, the inventors had admitted with great pride that Dolly was an exact replica of her sister. Well, added the court, not without irony, an exact replica of a farm animal is just another farm animal. You can't patent a run-of-the-mill farm animal, no matter that the method you used to make it was weird, even brilliant. Maybe you can patent the cloning method but not the sheep itself.

When the Dolly decision came down, I rushed to reread the patent on the Argentine cow. We never said that it was identical to a farm animal. We couldn't have said that, and we didn't because it wasn't. It was a human-made transgenic animal, which made any legal doubts go away.

## Waltzing Bionics

Since the Patent Office now agrees that genetically engineering an animal like BuenGen's cow is enough to make it capable of being patented, what if the animal is a human? We are, after all, members of the animal kingdom. Some may challenge that we are in good standing, since we seem to be intent on doing harm to other members, but no matter how obnoxious, members we are. And there is also no question that we have started modifying our bodies in ways that would make us an "article of manufacture" under the patent laws. Just think of folks walking among us with human-made skin grafts, orthopedic legs, or genetically modified pig hearts. Such bionic specimens are not natural and, I would argue, are sufficiently removed from nature to make them eligible. And, doubtless, in the era of gene editing that is now opening with CRISPR, sooner or later we will have genetically engineered human beings among us. Are we patent eligible?

Rest assured, we are not. The answer is found in a long-standing interpretation of the law by the Patent Office, which expressly prohibits obtaining patents "directed to or encompassing a human organism."[4] This interpretation has been described as inspired by the Thirteenth Amendment to the US Constitution: "Neither slavery nor involuntary servitude . . . shall exist within the United States." Owning human beings has been a constitutional violation since 1864. The Patent Office has extended that fundamental concept to intellectual property. It will not allow any patent on an animal if the patent claims are so broad as to include human beings.

Let's take a closer look at BuenGen's patent. Its main claim on the cow starts: "A non-human transgenic mammal . . ." Note the first few words, "A non-human." Had we filed a patent application with a claim that simply said, "A transgenic mammal . . ." the Patent Office would have forced us to qualify that the mammal was not a human. So, if you file an application on a new artificial hip, the language of the patent must avoid being so broad as to include a human waltzing around with it. You can patent a prosthetic hip in isolation from a human body, but if there is even a hint that the patent would encompass a person, it will be rejected until you correct it. Our society does not fancy seeing a human being with a patented prosthetic as a living, walking, and dancing infringement.

## Down on the Pharm

A few years after getting the patent for the transgenic cow, I found myself in Buenos Aires on a lecture assignment. I had the chance of meeting Dr. Barañao, one of the cow's co-inventors. He had since become the Argentine minister of science and technology and in that role tried—not always successfully—to strengthen the patent system in his country. Since I was visiting, BuenGen invited me to go see the patented bovines. The cows were not living in the company's headquarters in that residential neighborhood near the city center; that would have been over-the-top strange. The cows lived a few hours north of BA in a large farm between BA and

Rosario, the then second-largest city in Argentina. We drove out one early morning with the promise that, later that day, the vets were going to deliver four brand new transgenic calves by C-section. "Wouldn't that be interesting?" asked my clients. Never having seen a calf being born, never mind one I had patented, I readily agreed. When we arrived at the farm, one of my hosts, whose English was perfect, said to me in Spanish, "Nosotros llamamos a este lugar una 'pharm,' no una 'farm' " (We don't call this place a "farm" but a "pharm"). I laughed on cue, but I thought the joke was unexceptional.

Upon arrival, we hurried in since the C-sections were not waiting for us. We walked into a large open barn. There was an interesting assortment of veterinarians, gauchos, helpers, animals, and onlookers. The place smelled of a ranch, part fresh straw, bovine sweat, and old dung. It was far from an aseptic and sterile human operating room. There was a lot of hay, and a black pregnant cow was lying on its side, kicking at random times. The gauchos cautiously stayed away from her legs, until they tied her down. This "mother" was an Aberdeen Angus, a breed whose cows are good surrogates for embryo implantation and development. The vet operated on the mother, giving her a large horizontal cut. He and two gauchos pulled with all their strength at the newborn inside until she was out of there.

What I saw next blew my mind. What came out was a cute, tiny, black-and-white Holstein calf. The gauchos carefully lifted her off the surrogate's belly and placed her on a large rough table. There they cleaned her, made sure that she was breathing well, and carried her to a pile of straw, where they laid her down, and watched her for a while. Holstein cows are well known for their great milking ability. That's why this breed is ideal for genetic engineering of human proteins that will appear in their plentiful milk. But Holsteins are not good surrogates, and so their embryos are implanted into Angus mothers. Yet, a black-and-white Holstein born from an all-black Angus was as weird an event as I've ever seen. The only thing stranger would have been for the newborn to have a mark on its hide that said, "US Patent No. 7,807,862."

Figure 16.1, shows, on the left, the Holstein transgenic calf and her

Figure 16.1. *Left side,* veterinarian with Holstein newborn transgenic calf; *right side,* surrogate Angus mothers. Used with permission from Thomson Reuters.

vet, shortly after her birth; and, on the right, three Angus surrogates, one, her mother.

After the birth, the chief veterinarian sewed up the mother Angus, the gauchos loosened the ropes, and she shot up and trotted away without as much as a whimper or pained looked on her face. The vet, a gentle-looking man with white hair, started taking off his scrubs and turned to me. As his hat came off, I saw a long thin streak of bovine blood leading from his hairline to his eyebrows. He was clearly pleased with his labor. I asked if the Angus had been anesthetized. "No," he said, smiling. "They don't feel any pain." I wasn't sure I believed him. I certainly felt for that surrogate who had withstood a C-section without painkillers. But the black cow walked nonchalantly toward a group of three other pregnant Angus, who were about to be treated to the same procedure.

By the end of the morning the vet and his helpers had brought four tiny Holstein calves into the world. One of the four newborns died a few minutes after birth. As I lamented its death, I was taken aback by the response of one of my hosts: "That was one expensive calf; about a quarter of a million dollars," he said. "That's why we don't let them give birth out there in the middle of the night by themselves."

A little later, the whole team invited me to a classic Argentine asado, a *parrillada,* as we natives call it. It is made on a large low-lying grilling lattice covered with sausages, sweetbreads, kidneys, and a half dozen different cuts of the best beef the world knows. I half-jokingly asked if all that meat came from some patented Holstein cow, and they assured me that, of course, it didn't. The best beef comes from Angus, they said; Holsteins are milking cows.

But by then I had lost my appetite.

# Part IV

# Maintaining a Careful Balance

The biotech world—a term I use to broadly include universities or companies that play in the pharma, biopharma, or agricultural spaces—likes patents. It will go to great lengths to create exclusivity and maintain it for as long as possible. But not everyone thinks that exclusivity is beneficial. Other industries, like information technology (IT), which includes those who play in the electronics or software spaces, don't like it so much. And yet others, like diagnostic laboratories, are more nuanced in their views; they sometimes like patents and sometimes they don't.

Additionally, several US and worldwide constituencies blame the patent system for exacerbating problems of health care and food security. We have all read about high prices and poor availability of patented drugs, lack of access to the latest patented mRNA vaccines for the COVID-19 pandemic, and the role that patented GMOs have played in the loss of biodiversity. These dissenting voices are important; they warn us that if left unchecked, our patent system may become less controlled than originally envisioned. Let's explore whether the careful balance between the needs of research entities, investment communities, and the public that benefits from biopharmaceutical breakthroughs has tilted too much one way or the other.

# 17

# One Size Does Not Fit All

I first heard the dissenting voices on patents held by diagnostic laboratories in 2007, when I received a call from Dr. Marc Grodman, then the CEO of BioReference Labs, a clinical diagnostic company in New Jersey. Grodman was a classic hybrid from the age of commercial biology: a physician, professor, and corporate CEO. He was also a lively New Yorker with a sharp intellect, a good sense of humor, a well-trimmed beard, and a corporate jet. He practiced internal medicine in New York, taught clinical medicine at Columbia University, and, most proudly, led BioRef as its chairman and CEO. He believed that the future belonged to precision medicine, a modern specialty that uses patients' genetic constitutions to customize their treatment. Grodman wanted BioRef and GeneDx, a Maryland company BioRef had recently acquired, to lead the future with their highly specialized genetic diagnostic tests.

Grodman explained that he had patent troubles. His problem was with a bundle of patents held by a major competitor on the testing of genes affecting a cardiac condition called the "Long QT Syndrome" (LQTS). He wanted to challenge the patents or at least shake up the commercial arrangements.

"You know, we're trying to offer genetic diagnostics tests," said Grodman, "but there seem to be a lot of patents out there. I always knew that, sooner or later, I'd have to deal with patents . . . Ugh!" He appeared annoyed by patents, although his tone was more philosophical than irritated. He became animated as he started explaining the genetics of the Long QT Syndrome. I, in turn, became immediately intrigued by why an entrepreneur in the biotech industry could be annoyed by patents.

## R in One Place, D in Another

It is apparent to me now that the diverse views of exclusivity held by the IT and biotech worlds and, within the latter, the diagnostic labs arise out of their different business models. Most of their differing views have to do with risk: how to deal with it and how to hedge against it.

Several factors come into play when a company, be it a biotech or an IT one, wishes to put an innovative product on the market. Whether they want to launch a drug or a digital camera, all companies must invest on a few different fronts. Some of these are common between the biotech and IT industries; some are not.

The common areas have to do with paying for the fundamental research that may lead to a novel drug or integrated chip. In the last few decades, many large pharma and biopharma companies, let's call them Big Pharma, have stepped away from basic research.[1] They have left that field to universities or to smaller research-based biotech companies. If research is done by others, especially if paid for by the US government, big pharma companies can come in later and, instead of choosing among candidate drugs, pick and choose among different university labs or start-ups. The choice of a prolific university lab or small research company is a better long-term hedge against risk than the selection of one drug candidate over another.

I admit to simplifying, but, generally, small research groups, whether in academia or industry, tend to be nimbler and more innovative than those in the bigger, more hierarchical, and bureaucratic companies. Funding Research, the "R" in R&D, at a productive lab gives Big Pharma the possibility of a platform that will generate a steady stream of new inventions, which they send into their pipelines of drug Development. That's the "D" in R&D. Big Pharma then focus on using the platform for finding a new drug and financing the clinical work and marketing needed to turn a candidate drug into a commercial one.[2]

The biotech model of splitting R from D is neither limited to pharma nor to universities. You may recall from chapter 15 that it was not Monsanto's scientists who invented Roundup-Ready soybeans; it was researchers at Agracetus, a small California company doing basic work in plant genetics. And you may also recall that first Monsanto and ultimately Bayer ended up with the patent rights and with Agracetus. And the model of R in one place and D in another is not unique to the biotech world either. Big IT companies have long relied on research centers such as Stanford Research Institute or small software developers to provide them with a steady stream of new inventions and technological breakthroughs. The nimbleness of small companies works equally well in IT as it does in biotech. Big IT, just like Big Pharma, find a promising lab and, if it is a small startup, buys it.

Thus, both Big Pharma and Big IT march down the same early path of basic research. But then things diverge, and that's due to a few factors. The first is the amount of regulation biotech must navigate to put a product on the market. Because of the health risks involved in selling new drugs or genetically engineered food to the public, it takes years of expensive clinical or field trials to get approval from the FDA or USDA to do so. There are no such lengthy trials for the sale of a newly invented semiconductor. Since there are no public health risks with transistors, no expensive regulatory approval is needed for their sale.

A second difference between the biotech and IT worlds is in the lifespan of their products. A successful drug might be in pharmacies for decades. An integrated circuit that controls the color intensity of a computer screen may outlive its attractiveness after a few years, if that long. If it takes an electronics company two to three years to get a patent, the product it was meant to protect may well have run its course in the marketplace. Things move more slowly and with much more cash burn in the biotech world than in the IT world. That's why patents are so important in the biotech world.

Yet a third difference between the industries—at least in the past—was the existence of "patent thickets" in the world of IT. A patent thicket has been defined as "a dense web of overlapping intellectual property rights that a company must hack its way through in order to actually commercialize new technology."[3] The thicket of patents that covers an electronic product such as a laptop computer or a digital camera is oftentimes impassable. I don't mean only the patents held by the company that makes and sells the devices. I include the many patents held by others on individual components: circuits, optics, software, materials, and designs. It has been estimated that the maker of a complex consumer device, such as Apple or Nikon, may need hundreds of licenses from other patent holders before it can sell a new type of laptop or camera.[4]

Historically, the number of patents per average pharmaceutical drug used to be only between two and four.[5] However, things have changed dramatically since the advent of biological drugs, a topic on which I will have more to say in chapter 19. The powerful antirheumatic drug Humira, made and sold by AbbVie, is estimated to be covered by close to 130 patents.[6] Some patents are on the drug, some on its formulations, some on dosages, and many on methods of manufacturing or purifying it. That is a thicket as impenetrable as that for a new laptop computer. The main difference with a computer, however, is that the patent thicket is on a single drug, Humira, and all the patents are owned or controlled by AbbVie, a single company. A company wishing to sell a competitor drug is still blocked by AbbVie's bundle of patents but is not facing 130 patents held by dozens of different companies.

The IT companies have a love-hate relationship with patents. Because of the many patents held by their competitors, they will apply for and get plenty themselves, so when the moment comes, they can horse-trade. If it looks like the river of innovation is about to seize up and freeze due to the many floes moving downstream, it's time to do deals. Or as is the custom in the IT industry, they put all needed patents into a so-called patent pool, and let everyone have access for a fee. Otherwise, no one sells anything. If it weren't for the thickets of patents owned by multiple owners in the IT industry, I think that most of the companies would much rather not deal with the patent world. Or at least they'd get many fewer patents than they do. Not so in the biopharma world, where collaborative mechanisms such as patent pools are rare to nonexistent. The culture of the biotech world is single-lane swimming, not mingling in the pool.[7]

Now, regardless of the speed or cost of R&D, or the many patents held by multiple competitors, no one likes to have their products pirated by bad players, whether the products are biological cells or electronic circuits. But because the amount of investment and the risk of failure involved for living cells are much higher than for circuits, the pain of piracy bites harder in biotech. This fundamental difference between the industries has led to a rift between them as to the value of patents. They don't see things the same way. When there are public debates on changing the patent laws to incentivize investments and bring more innovation to the public, different US patent constituencies have differing responses.[8] The split follows the risk, or lack of it, involved at different stages of a product's life: basic research, lengthy preliminary trials, regulatory burdens, and the possibility of thickets of patents controlled by multiple different owners.

The IT world is rarely interested in making it easier to file for and get patents. Except as bargaining chips, IT companies don't need them as much; their products may lose market dominance before patents even issue; they are perfectly happy to compete with others on the weight of their marketing savvy and on their ability to move on to "the next big thing" coming down the pike. They will go for many patents of narrow scope, rather than just a few but of broader scope.

The drug makers generally are at the other end of the spectrum. They need a few strong patents of wide scope to defend their beachheads against pirates, especially during many years of uncertain research and expensive clinical trials. Their products will live on for decades, so they want as lengthy a patent life as they can lobby the legislators to give them. And while they may see the "next big thing" over the horizon, they know that it will not be easy to get there.

The conclusion is that a "one size fits all" patent system like ours has caused a split between the industries. There are no separate patents for electronics or software and for biotech; we have one set of patent laws. Yet the statutory sameness has led to a political and legal gap among constituencies with different business models, such as those in the biotech world and those in the IT world.

I learned from Marc Grodman that the diagnostic companies are sometimes in the middle.

## Energetic Athletes

Within the biotech world, there is a difference between pharmaceutical companies that develop and sell drugs, such as Biogen or Amgen, and diagnostic companies that develop and sell lab testing services, such as Quest Diagnostics or LabCorp. This split also has to do with the nature and amount of regulation. Diagnostic labs, which offer clinical analyses for a fee, are not as tightly regulated by the FDA as biotech companies that sell drugs. Therefore, patent exclusivity to incentivize investors matters less to the labs than to the drug makers.

In his 2007 call, Grodman made it clear that when it came to genes used for diagnostic procedures, a field aptly named "genetic diagnostics," he was not enamored of gene patents and would have rather that they didn't exist altogether. As you may recall, in 2007 I was a strong believer in patents on isolated genes, supported in my views by Judge Alan Lourie of the Court of Appeals. It turns out that BioRef, which was intent on becoming a pioneer in the emerging field of "personalized" or "precision medicine," had different ideas.

The concept of personalized medicine is to stratify a patient population into groups in accordance with their genetic constitutions. A doctor determines to which group an individual patient belongs and treats him accordingly. A clinical lab like GeneDx will look for the presence or absence of a known genetic mutation in an individual patient and will inform the attending physician whether the mutation is present or not. If the mutation is there, the doc, in response, will administer a drug; but, if not there, she will not administer the drug, or administer a different drug.

Figure 17.1. shows the basic idea behind personalized medicine.

The figure has four columns, as seen from left to right. The leftmost portion shows a group of patients with the same disease (left column), which, using genetic diagnostics (double helix in second column) is stratified into three subgroups (third column). Depending on their genetic constitutions, two of the subgroups receive their own specific medications, and the third subgroup receives none (right column).

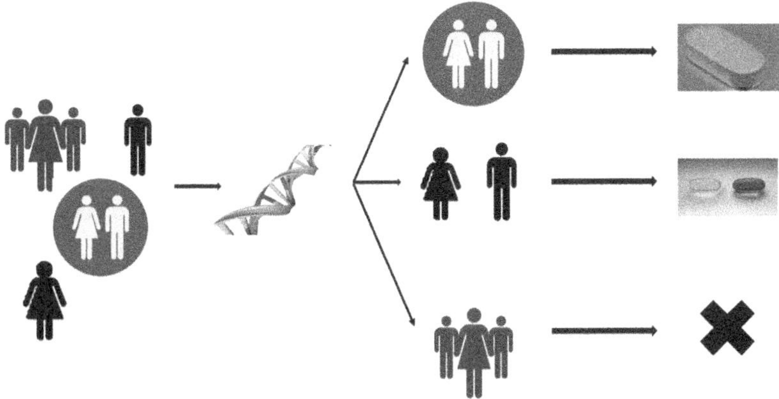

Figure 17.1. The basic concept of precision medicine. All components in the figure are from Wikimedia Commons and are either in the public domain or are licensed under the Creative Commons Attribution-Share Alike 4.0 International license.

"My immediate problem," said Grodman during the initial call, "is with the patents on the Long QT Syndrome."

As he explained what LQTS was, I realized that while I had never heard the name of the syndrome, I sure had heard of energetic athletes, such as basketball players or runners, suddenly dropping dead on the court or field. Jim Fixx, the jogger, and author of *The Complete Book of Running*, died of a sudden cardiac arrest in 1984 at age fifty-two, while on his daily run. Reggie Lewis, the Boston Celtics basketball player and one of National Basketball Association's leading scorers, died in 1993 at age twenty-seven, also from a sudden heart attack. In 2021, years after my introduction to LQT Syndrome, Denmark's most famous soccer player, midfielder Christian Eriksen, had a sudden infarct at age twenty-nine during an international game. He survived, thanks to quick medical intervention.

What these athletes have in common is Long QT Syndrome, an arrhythmia that can be diagnosed by the length of the so-called "QT" interval on an electrocardiogram. The congenital version of the syndrome is caused by genetic abnormalities. If you can detect the gene mutations involved with the syndrome, you can diagnose risk and act accordingly. The *medical* complication is that there were in 2007 no less than twelve identified genes associated with different clinical variants of the syndrome; they were named LQT1 to LQT12. Let's say that a young patient came to you with severe deafness in both ears. If you ran genetic diagnostic tests on the group of twelve LQTS genes, you might discover that the patient had mutations on LQT1 and LQT5. If so, she would be likely to also have an arrhythmia that put her at risk of a sudden heart attack. Look back at figure 17.1, and imagine that instead of stratifying a patient population with LQTS into three groups, you stratify it into twelve groups, each one associated with a different gene. If you were able to do so, you could go beyond the naming of the syndrome and could understand the exact reason and the expected clinical consequences.

The *legal* complication in 2007 was that if you wanted to set up a diagnostic panel including tests for all twelve genes, you had to face at least one patent for each gene. In fact, there were more than twelve patents out there. There were patents on the isolated genes in their natural forms, in their mutated forms, and on methods of genetic diagnostics using the correlation between genes and clinical conditions. The problem for Grodman at the time was that all the patents on the LQTS genes were controlled by one company, Clinical Data, Inc. (CDI), of Newton, Massachusetts. The patents, owned by the University of Utah Research Foundation, the commercializing arm of the University of Utah, were the products of groundbreaking research on the genetics of LQTS, by Professor Mark Keating. Utah, in turn, had licensed them exclusively to CDI.[9] The dozen or more of these patents formed what looked like a crenelated fortress with twelve towers, all of them protecting CDI's complete exclusivity.

After telling me this story, Grodman said that he had gone to other law firms for assistance. Most of them had told him that gene patents were here to stay, and they couldn't do much to help him. Grodman then asked me, "Can you help us? I'm told that you are a specialist in gene patents, and that you've gotten quite a few for your clients. I think that gene patents and correlation patents are hurting the diagnostics business.

So, even though you have supported gene patents, will you still be able to represent us?"

I was ready for this case. I had been studying patent fortresses in the gene diagnostics industry for some time.[10] I had speculated that if the overall exclusivity could be fragmented so that one entity controlled some patents and another controlled other patents, the fortress would transform into a thicket with multiple owners. Each entity would own a piece of the whole and couldn't commercialize the complete set without the agreement of the other. Since the partial owners blocked each other, they would have to sit down and negotiate. Maybe, I thought, they could then form a genetics patent pool, just like the ones that were common in the IT patent thickets. I had been anticipating a case like the one that Grodman was bringing to me.

"Yes," I replied. "I've been waiting for this challenge for some time. Happy to help."

"Good!" said Grodman, invigorated. "Let's go to battle!"

Battle. That was music to a lawyer's ears.

I explained that there were two live cases pending before the courts dealing with these issues. One or both might end up at the Supreme Court. The patents of Myriad Genetics on isolated genes were being challenged in New York by the Association for Molecular Pathology (AMP), as claiming subject matter that was not eligible. I, like so many in the biotech patent community, thought that a ruling that would get rid of patents on isolated genes, while solving Grodman's problems, was a distant—even unlikely—possibility. (As you may recall from chapter 10, however, I and other skeptics were ultimately proven wrong.)

The other case was a fight in California between the Mayo Clinic and Prometheus Labs over the validity of patents on so-called natural correlations. Prometheus had patents on correlations between drug dosages and effectiveness,[11] and they were being challenged by the Mayo Clinic as being nothing more than attempts at patenting a phenomenon of nature. Patents on similar correlations, like those between the twelve LQT genes and the LQT Syndrome, had grown like weeds in the previous decade. By 2007 there were so many patents on genetic correlations that the field started looking like the thickets so common in the IT world. If the court ruled in favor of the Mayo Clinic, it might also help Grodman.

## Red Sky Correlation

What are correlation patents? Let me try and explain them using a weather example. In the process of doing so, I will also explain why in 2007 the underlying diagnostic inventions were being challenged by the Mayo Clinic as not patentable.

Old folklore has it that "red sky at night, sailor's delight." We know today that when the western sky is red at sunset, high-pressure air and good weather are moving eastward and heading our way. That is a very useful approach to forecast the weather and to plan accordingly. Imagine the cavewoman who first figured out the relation between red skies and the next day's weather. Looking at the sunset, she may have said, "Tomorrow is a perfect day to go hunting; let's all get up early." Her insightful prediction may not have been that obvious to the rest of her band. Yet, because it was nothing but observation of a natural phenomenon, it would not have been eligible for a patent. A discovery may be brilliant and useful yet not be something that can be patented. In today's legal parlance, "non-obviousness" and "lack of eligibility" are different concepts. The discovery of a correlation between 1 of the 20,000 to 30,000 genes in the human genome and some aspect of the LQT Syndrome may also be brilliant and useful. However, just like the red sky–good weather connection, it may not be eligible for patents since it is nothing but a natural phenomenon. Whether the discovery of a correlation between drug dosages and effectiveness was eligible or not was the question being asked in the Mayo Clinic litigation.

I explained to Dr. Grodman that the Mayo Clinic case might wipe out many patents on genetic diagnostic correlations, including some controlled by CDI. And the Myriad Genetics case may wipe out patents on isolated genes. But, I added, the resolutions of either the Myriad Genetics or the Mayo Clinic cases were uncertain and in the future.

"Yeah, I know that getting rid of these kinds of patents is an iffy proposition," said Grodman. "I don't want to wait. I want to have hearings on the Hill on this business and I want to testify," he continued. "I have asked my friend Elizabeth Holtzman to help with this. You know her, no?"

I had heard of Holtzman: member of the House of Representatives from New York; a seat on the Judiciary Committee, which had recommended impeaching Richard Nixon; district attorney for Brooklyn; and New York City comptroller. Most recently, she was a prominent and well-connected New York lawyer. "I'd be happy to work with her," I said.

"This will be fun!" said Grodman. The man loved to mix it up.

Holtzman and I soon became buddies in Grodman's battle. We would plan strategy. We would exchange scripts for Grodman's testimony, and we would keep each other honest in our respective areas of expertise: hers, legislative insight; mine, biotech patent law.

The high point of our efforts was a hearing on October 30, 2007, before the House Judiciary Subcommittee on Courts, the Internet, and Intellectual Property. The hearing was entitled "Stifling or Stimulating—the Role of Gene Patents in Research and Genetic Testing." Grodman was one of the main witnesses, and he testified against exclusivity. It was, as I mentioned earlier, through preparing and listening to Grodman that I realized that not everyone in the biotech world saw patents the way I did: that they're always a good thing. Not so, said Grodman in his testimony:

> There is a fundamental difference between the situation of drug companies and diagnostic laboratories. Given the huge costs of drug development, denying patent rights . . . could have serious consequences for the willingness of companies to undertake the needed research in the first place. Without solid patent protection, the companies could see no way to recoup their enormous investment. But in the area of gene patents for diagnostic tests, efforts to identify new genes and their correlation with disease would not seriously be discouraged by the absence of exclusive patent rights. . . . The costs

> of discovery [in genetic diagnostics] are not comparable to those for drug development. . . . [I am] in favor of a regime where a company like mine can obtain a non-exclusive license from the holder of the patent. . . . If I can demonstrate that my test would be better, faster, provide fewer false negatives or positives, fill a niche, cost less to the public or perhaps complement the test already offered by my competitor then the public will benefit greatly by my entry.[12]

He ended his testimony with a quotable flourish:

> I am not asking for a free ride; all I am asking for is the ability to compete fairly and benefit the public and my company. In the area of genetic testing, exclusivity is a formula for mediocrity.[13]

"Oh, my," I thought, as I heard him testify, "exclusivity is a formula for mediocrity." I was not the one who put those fighting words in Grodman's mouth. Those were his words. I couldn't go that far.

A lawyer representing the Biotechnology Industry Organization pushed back, vehemently disagreeing with Grodman. He testified that any watering down of patent rights would undermine investments in research and development. His was the basic "all-patents-are-good" position, which was my view before getting involved with BioRef. No subtle differences; no nuance.

## Cracking the Fortresses

After the hearing, Congress, as is its wont, did nothing about the diagnostics controversy. Frustrated with waiting for Congress or for the courts to address the practice of patenting isolated genes or genetic correlations, Grodman, in his inimitable fighting style, renewed the battle. He discovered that two of the exclusive licenses from Utah to Clinical Data Inc. had not been renewed. To this day I don't know for sure how he found out. I suspect that someone with inside information tipped him off, but that is pure conjecture on my part. In the event, Grodman got on his jet, flew to Salt Lake City, and secured the two exclusive licenses that CDI had allowed to lapse. He now controlled a piece of the whole for the testing of LQT Syndrome; some was his, some was CDI's.

Grodman had cracked the fortress of solely owned exclusivity and turned it into a blocked thicket with dual owners: CDI and GeneDx. These two were as blocked as the patent owners we met in chapter 15 where one had a patent on TV and the other had a patent on color TV. Neither of them could sell color TVs without a cross-license from the other. Same here: neither CDI nor GeneDx could offer a full diagnostic test for the twelve LQTS genes without getting a license from the other.

Grodman's move to create a dual-owned LQTS patent thicket received strong criticism from Drew Fromkin, the CEO of CDI. Fromkin publicly implied that Grodman was being hypocritical. It seemed to Fromkin that Grodman did not support exclusivity in the diagnostics industry unless the exclusive rights were his. Grodman's answer was that his exclusivity was strategic. He wouldn't sue anyone; all he wanted to do was to break up the fortress. Professor Robert Cook-Deegan of Duke University, together with Misha Angrist, two of the coauthors of a 2010 study of the patent wars surrounding LQT Syndrome, seemed to agree with Grodman. In a seminal article on LQTS patents, they said, "Securing an exclusive license is . . . not necessarily hypocritical, if it is a strategy to induce negotiation in the face of existing exclusive rights."[14]

The result of the newly created multi-owned thicket was a standoff. No one sued anyone. Both CDI and GeneDx offered full panels of gene testing for LQTS, each including all necessary genes. Both companies watched each other across the logjam, but no one crossed the river to negotiate. In the end they just moved on.

Grodman was one creative and gutsy client. *Chapeau.*

The other patent fortress Grodman and I took on during those years had to do with the *BRCA* genes affecting breast cancer. GeneDx would determine mutations in the *BRCA1* and *BRCA2* cancer genes and identify any given patient as belonging to one of two groups. If the genes were mutated, the patient belonged to group one, and the results would suggest a higher risk of breast cancer. That, in turn, would lead to different treatment decisions than if the genes were not mutated. In such case, the group-two patient might not be treated at all.

You may recall from chapter 10 that these *BRCA1/2* genes were involved in the landmark case of *Myriad Genetics,* which went up to the Supreme Court in 2013. The Court decided then that isolated genes or even fragments of isolated genes were not subject matter capable of being patented. The Supremes held that the basic function of genes, whether embedded in the genome or in a test tube, was to carry information. No new properties or functions were made available after isolation than before. Because they were not legally far enough from the identical genes as they appear in nature, you could not patent isolated genes, regardless of how pure they were. This was good news for Grodman and BioRef.

The other good news for our clients had come a year earlier in the Mayo Clinic litigation. In its 2012 decision in *Mayo v. Prometheus,*[15] the Supreme Court held that patents on natural correlations, such as those between gene mutations and disease, were nothing but "laws of nature," and you couldn't patent laws of nature. The correlation between a mutation in the *BRCA1* gene and breast cancer is not an eligible invention. As the relation between red skies and good weather, it is nothing but a natural phenomenon.

In both the *Myriad* and *Mayo* decisions, the Supreme Court made it clear that no matter how brilliant the discovery, if a patent claim got too close to a natural product, like a gene, or to a natural phenomenon, like the correlation between a gene mutation and disease, it would step over the line of permissible eligibility. Both types of patents would be knocked down in court. Like Icarus, the two patent holders, Myriad Genetics and Prometheus Labs, had flown too close to the sun, and their wings melted. The IP gods had aligned for Grodman.

In late 2013, reassured by both decisions, BioRef started commercializing the use of isolated *BRCA1/2* genes or their fragments in testing protocols, the details of which were posted on their website. Notwithstanding the legal debacle for isolated gene and correlation patents, Myriad Genetics did not give up. The company had thirteen more *BRCA1/2* gene patents in its portfolio, which had not been directly involved in the litigation it had just lost. These patents were similar but not identical to the ones the Supreme Court had vanquished. Some of them included patent claims to *BRCA1/2* in cDNA form, which had been held eligible by the Supreme Court. Yet to any objective observer, they were the walking dead. Seemingly oblivious to the loss it had just suffered, Myriad sued BioRef and six other companies for infringement of the thirteen zombie patents in Federal District Court in Salt Lake City, Utah. The reasons it did that escaped me then, and still do. It felt like harassment.

When Grodman got served with the complaint, he called me. He was irate. "We've been sued by Myriad! Do you believe it? The chutzpah!"

It was back to the battering rams to counterattack Myriad's patent fortress. But the arrows in the quiver of Myriad Genetics were by then severely blunted. About a year after starting its new battle, in early 2015, we got a call from Myriad offering settlement. It is a very rare case when someone who starts a fight vies for peace after twelve months. It suggests that whoever was in denial over at Myriad was overruled by someone with a cooler head. And that was the end of the gene patent wars.

*Myriad* and *Mayo* unblocked the IP fields of genetic diagnostics and of precision medicine. After that, BioRef and GeneDx were able to move forward with their plans to become pioneers in personalized medicine, unhindered by patent fortresses, thickets, or lawsuits. I learned that even in the life sciences world, our "one size fits all" patent system is not as homogeneous as I once thought.

It may be useful for an educated consumer, constantly bombarded by public pronouncements about all sorts of topics, to understand the patent split between the biotech and IT worlds, and between the drug makers and the diagnostic labs. When some spokesperson warns that "without patents we'll stop being an advanced society," and a dissenting one decries that "patents are monopolies, and monopolies are a terrible thing," ask yourself: Where is this coming from? Remember, not everyone in our one-size-fits-all patent system sees exclusivity the same way.

# 18

# Controlled Monopolies

The questions asked by the dissenting spokesperson in the previous chapter are worth further inquiry: Are patents monopolies? And, if so, are they terrible?

The law professors who taught me intellectual property in the early 1980s would routinely insist that patents were not really monopolies. They would explain that while economic monopolies were bad, a modern patent was not an economic monopoly. A monopoly is a situation where a company has enough market dominance in some area of the economy to manipulate prices and drive out competition. Such monopolies may be encouraged by a corrupt government or may be achieved by rapacious corporate behavior. "But," the law professors would say, "contemporary patents don't come from either of those. At best, they're controlled monopolies."

"OK," I would think, "but what happens when things get out of control?" It seems to me that in the present era, the original Venetian concept of patents has become distorted. Two such distortions have arisen: one, involving patented drugs, has had an impact on prices, and the other, involving patented crops, has had an impact on the world's ecology. In this chapter and the next, we'll talk about drugs and crops.

## High-Priced Drugs

We live in an era of unprecedented inventions and improvements in pharmacology. There are constant innovations in drugs for all kinds of previously incurable diseases, such as intractable cancers, autoimmune diseases, and even genetic defects. The companies that by themselves or in partnership with academia bring us these innovations are the poster children for supporting exclusivity. They operate in a highly regulated and very risky health-care market. They incur enormous expenditures per drug. The patent system deserves a large amount of credit for bringing us such wondrous pharmaceutical developments.

Drug research is a high-risk proposition. Not only do scientists need to find a molecular structure that shows some beneficial effect in a lab animal, such as a rat or rabbit, but they then need to be able to turn it into a molecule that can be administered to a human, by injection, infusion, or, ideally, by mouth. This is known in the jargon as making a promising molecule "druggable" for humans. At this point in the process, the hope is that the drug will show the same effect on the human body that it does in animals. Next, the scientists need to figure out the right human dosage to maximize effectiveness while minimizing nasty side effects. After this, the drug must go through extensive clinical trials to demonstrate that its effectiveness outweighs its toxicity in large human populations. The hope is that with good results, the FDA will agree that the risk-benefit balance leads to a drug worthy of approval for humans. And only then is the drug commercialized.

The road is long, full of obstacles, and very expensive. That's where patents come in. When a researcher makes a test tube discovery of some sort of bioactive molecule, she runs to her patent department, which quickly prepares and files initial applications at the Patent Office. The lawyers try the best they can to protect not only the new structure but also related structures, so that competitors will not be able to design similar molecules while easily avoiding the initial one. It is not unusual that after a few years of trial and error, including modifications of the original structure, the scientists discover that the ultimately best molecule was not specifically identified in the first patent filings. So, a new set of filings on the best candidate, the "lead" as it is known, get filed. The original filings still block access to the family of similar ones, and the subsequent filing protects the specific commercial product. Both sets of filings start to create a fence around the invention. This iterative patent process leads to patents on the best structure, best dosage, best formulation, best mode of administration, and best methods of manufacturing. It results in multiple patents surrounding the drug, especially if, on the road from test tube to humans, it has gone from being a promising lead to becoming a blockbuster. This is the "bundle" of patents owned by one entity that we discussed in the previous chapter.

Of course, extensive patent protection leads the innovator company to be able to charge high prices for the successful drug, especially if the drug has been a breakthrough and has no competition in its medical field. Part, albeit not all, of the high price of a patented drug goes toward recouping the investments made at the different stages of research and development for the drug at hand. In addition, it also funds the investments made into the myriad of disappointing candidates that have failed the R&D process. And part of the high price funds the costs of marketing and commercialization. In turn, the high charges for new medicaments give rise to complaints that innovative patented drugs are more expensive than justified. The pharma industry argues that prices need to be high because doing drug research is expensive. The critics argue that the process of setting prices for patented drugs is "monopolistic." Who is right? Are high prices the fault of the patent system? My answer is: to some extent but not entirely.

Let me first put some perspective on the common complaint that drug patents lead to "monopolies." To do so, it may be useful to provide some background on monopolies, starting with the early, corrupt, and rapacious ones. A classic example of a corrupt government monopoly is that of Queen Elizabeth I. In the sixteenth century, Good Queen Bess gave so-called patent monopolies to a select few of her favorite subjects on regular things like salt or playing cards. In turn, her friends were willing to hand over large fees and enrich the Crown's coffers. Only her friends could sell these common articles; no others could. In 1624 the British Parliament ended this shady practice by enacting the "Statute of Monopolies." It made all monopolies illegal, *except* those for new and creative inventions, and then only for a limited period. This very exemption carved out from the Statute of Monopolies in the seventeenth century became the basis of today's modern patent systems.

An example of a monopoly brought about by rapacious corporate behavior is Rockefeller's Standard Oil. During the late nineteenth and early twentieth centuries, the company monopolized the US supply of oil. It kept prices low for a while, and, once its competitors were gone or had been gobbled up at fire sales, the company drove prices up. As Standard Oil's control of the market increased, the government intervened, filed antitrust actions, and broke up the company in 1911.

"You see," my law professors would continue, "a modern patent of invention has nothing in common with the corrupt monopolies of Elizabeth I or with the rapacious monopolies of Standard Oil. Nothing is being taken from the public that it had before the invention came on the scene. There are no patents on crude oil, playing cards, or salt. No one is entitled to get exclusivity on things that are in the public domain or its penumbra. In addition, the patent holder will usually not have the kind of market dominance of a Standard Oil and be able to destroy competition. A well-enacted patent system, based on giving exclusivity for a limited period and only for 'ingenious' inventions, does not create an economic monopoly."

I was skeptical, no matter how eloquent my professors. I had the feeling that when it came to contemporary patents, my professors hated the very word "monopoly." OK, I would reply, so a corrupt or rapacious monopolist drives out competition by manipulating prices and, once it dominates the market, is able to set any price it wishes. I got it. But isn't the holder of a lawfully obtained patent on a cancer drug, even if he is honest, allowed to set the price as high as he wants? And if so, aren't such prices "monopolistic," in the sense that there is no competition to bring them down?

My profs' answer was that the patent monopoly is only on the new cancer drug itself and any foreseeable variants covered by the patent. No one stops another smart scientist from simply "designing around" the patent and selling a similar drug outside its scope. Or even if that's not possible, there may be perfectly effective drugs out there, which may once have been patented but are now generic, and which give the patented drug some competition. In other words, there's no Queen Bess or John Rockefeller to control the market in drugs for cancer; whatever "monopoly" exists is drawn to the patented drug, its formulations, and similar members of its family. And the patented drug may be expensive for no more than about a decade and a half, after which it goes into the public domain and the prices drop.

The Socratic dialogue between my profs and me might end with something like "Now sit down, Goldstein." I understood, of course, and I would sit down. During the seventeen or eighteen years of actual patent exclusivity, no one else can sell the patented drug, which may be "the latest of the latest" innovation. For a while, the public may not get inexpensive access to the latest, but that's the price society is willing to pay for incentives to innovate. You may remember from chapter 1 that Thomas Jefferson, while calling this kind of temporal exclusivity the "embarrassment of an exclusive patent," was willing to tolerate it for the long-term good of society. And in the same chapter, I quoted Judge Randall Rader as calling the temporal exclusivity a "generational gift."

While I do believe that a well-run patent system is a terrific thing, the US practice has been subject to clever maneuvering. As I will show you, it has been possible, by perfectly legal manipulations, to prolong exclusivity well beyond the term of any one single patent. Consequently, at least in the pharmaceutical context, the costs and benefits of patents are often out of balance. The laissez-faire ideologues ignore these maneuvers and would just let the market solve the problems. When asked about high drug prices, they will say (usually invoking Jefferson), "Well, exclusivity gives the patent owner the right to charge whatever the market will bear. That's the price we all pay to promote innovation. It's for the common good that we have pharma patents."

I would answer, "Why do we define 'the common good' so narrowly as not to include drug affordability?" I am not an economist, nor do I pretend to understand pharmaceutical marketplaces very well. But in my view, the problem with high-priced drugs is more complex than patent maneuvering. Patents play a role, sure, but the problem goes beyond them. It also depends on the existence (or not) of a centrally coordinated health-care insurance system that would give an entity like Medicare the ability to negotiate transparent prices with a single voice. The balance of the marketplace can also be pushed one way or the other by the ups and downs of our government policies, which, depending on the prevailing wind, are at times hands off and at times engaged. And the proper direction of the wind is secured by the mighty pharmaceutical industry, which in terms of annual expenditures, is considered to have the number one lobbying group in the United States.[1] The whole system has the feel of a multicolored wind spinner, which, when first built, used to rotate at dizzying speed but has now been slowed down almost to a halt by entrenched economic interests.

## March-ins and Takings

You might ask: Isn't there some way to use the existing patent system itself to address the problem of high prices? The short answer is not really. The US government has two limited options to intervene in the marketplace of patented drugs: they go by the names of "march-ins" and "takings." And, the government has been reluctant to use them.

In chapter 3, I explained that in 1980 the Congress enacted the Bayh-Dole Act. The legislators tried to put some limits on the unrestricted rights of federally funded inventors. Under Bayh-Dole, the federal government has "march-in" rights that are meant to hang, like a sword of Damocles, over the recipient or its exclusive license holder. If the health or safety needs of the public remain unmet, the funding agency can take the underused patent rights away and give them to someone else.

As early as 1999 and as late as 2017, there have been attempts to get the NIH, the main funding agency for health sciences in the United States, to "march in" and take away the patent rights from a license holder of a patented drug who was alleged to be charging too much money for it. The argument went that if the price of a drug is too high, the health needs of the public remain unmet. None of the attempts succeeded. Regardless of the party in power, the NIH has never deviated from its view that it will not get involved in price disputes involving patented drugs made with federal funds.[2] By now, most recipients of federal funds and their investors have discounted the march-in provisions of Bayh-Dole as toothless. The refusal of the NIH to use march-in rights as a way to lower drug prices is the subject of intense disagreement between activists who favor the use of march-ins for price reasons,[3] and the pharmaceutical industry, which is strongly against march-ins.[4]

There is another mechanism in the patent law to try and ensure that patented inventions are available at lower prices but only for use by the government in cases of public need. In this situation, federal funds need not have been involved in the creation of the invention. The procedure goes by the legal name of "a taking."

The government has always been able to take private property for public use under the concept of "eminent domain." It will routinely take private land for public easements or for public buildings. And just as with land, the government can take patents. The only limit to this power is that the patent be taken for a "public use," that is, for the good of the nation. Under the Fifth Amendment of the Constitution, all a patent owner can get in return is "reasonable compensation," an amount decided in the US Court of Federal Claims.

I doubt that the government would threaten to "take" the patent rights to an expensive cancer drug that is used on a small portion of the population. That would not be "public" enough. A patent "taking" might be justified during a runaway pandemic such as COVID-19. I am sure that the pharma companies that produce vaccines—such as Moderna, Pfizer, or Johnson & Johnson (J&J)—know quite well that the government could invoke a taking of their patents and start making vaccines with a generic contractor of its choice. If the government did that, the companies would be left with suing the United States for "reasonable" compensation. Much better to do a deal with the government, sell them vaccine stocks at "reasonable" prices, and avoid litigation. In other words, the threat of a taking without *ever* taking is enough. It's a powerful tool in the negotiating armamentarium of the government.

The bottom line on lowering the high prices of patented drugs is that there aren't a lot of ways to bring them down through the patent laws. Until such time as government policy changes, the government will be reluctant to "march in" to lower the price of a patented drug invented with federal money. If the patented drug was invented with private funds, the government could "take" the patents but only if it is for use by the government in some sort of public emergency. The *threat* of a taking, however, is a constant reminder to private patent owners that when they are selling things like COVID vaccines, which can easily be described to a skeptical judge as critical for the public health of the nation, they better step lightly and not get too greedy.

Since both a march-in and taking of a patent are ineffectual ways of lowering the prices of patented drugs, the generic companies are left with the choice of "designing around" the patent, trying to invalidate it in court or in the Patent Office, or waiting for it to expire. And expire it will. Then, like hummingbirds flocking to a container of sugar water, the generics will come, and prices will drop.

And prices do drop but . . . things are never that simple.

## Submarines and Evergreens

The pharmaceutical drug world is a cutthroat business. The generic firms are constantly nipping at the heels of the companies that invent and sell top brand-name drugs; let's call them the "brand-name" companies. Both sides spend fortunes on patent attorneys to try and protect, or (if "on the other side of the 'v.'") to break, the exclusivity of a blockbuster. Large legal fees are small change when trying to hang on to or catch a slice of multibillion-dollar yearly markets. Over the years, the brand-name companies have used a few maneuvers to lengthen patent life. Some are no longer around, others still are.

One maneuver used in the past but no longer available was to keep a chain of patent applications pending for as long as possible—and do that in secret. This was advantageous to patent applicants because, before 1995, patents expired seventeen years from the day of *grant.* If a pharma company filed an application in 1990, and kept it in the Patent Office for ten years and it was granted in 2000, it would then expire in 2017. No one would know that something was pending until the day of grant, when, like a U-boat breaching the calm sea, a patent was suddenly flung upon unsuspecting competitors. It shouldn't surprise you to learn that such abruptly appearing patents were known by the outraged competition as "submarines." No one ever knew how many and what kinds of submarines were navigating stealthily inside the examination rooms of the Patent Office. Thank goodness this regulation was finally changed in 1995. Patents filed after that date expire twenty years from their *filing* dates, so our exemplary patent would expire in 2010, seven years earlier. There may still be a lone U-boat under the bureaucratic waters, but it is becoming less likely that it will surface any time soon. (But stay tuned for the story of the Enbrel submarine, which I will tell you in the next chapter.)

Another maneuver still actively used is extending the patent life by what is known as evergreening. Knowing the inevitable expiration fate of their valued patents, the brand-name companies have figured out ingenious ways to prolong exclusivity for their best-selling drugs. These (entirely legal) tricks of the trade that are used to prolong patent life go by the name of "managing the lifecycle." They are aimed at protecting the patent life of a lucrative drug, from discovery to advanced formulations given to patients. The more colorful word "evergreening" evokes spruces that never lose the color of their leaves, no matter the season. Pharma companies with a commercially successful drug that is patented will apply for additional patents on follow-on improvements. These include things like different formulations, different dosages, mixtures with other drugs, and slow-release modifications. If you scratch the surface of branded pharma patent estates, you will usually find that the whole enterprise of selling a blockbuster is surrounded by a thick picket fence, a thicket, which defends against generic copiers. No matter the expiration of the original patent, there are usually later ones to take its place in the evergreen fence.

Sometimes, however, the bag of legal tricks runs dry. There are only so many "non-obvious" follow-on improvements that can be patented after the basic drug and its uses have been discovered. Evergreening has its limits. In such situations, the brand-name patent holders may try a commercial move known in the vernacular as the "if-you-can't-beat 'em-join 'em" maneuver. This consists of selling generics *of their own drugs*, so-called authorized generics. A traditional generic is a copy of the brand-name drug that is sold by a company *other* than the one that holds the patent on the original drug. It can only be launched by the other company if the patent on the brand-name version has expired or has been legally invalidated. In contrast, an authorized generic drug can be launched by the brand-name company even while its own patents are still in place—and certainly afterward. There's no one there to stop them.

An example of this maneuver is the anti-inflammatory drug celecoxib. It was invented by the pharma company G.D. Searle, which filed for patents in 1993. Pfizer eventually acquired the rights to the drug and the patent and started selling it under the brand name Celebrex®. The drug went off patent in 2014, at which time the FDA approved Teva Pharmaceuticals of Israel to sell a generic version. Not to be left behind, however, Pfizer, the now patent-less seller of Celebrex, went into competition with Teva. Through its subsidiary Upjohn (now Viatris), Pfizer kept selling the drug as an authorized generic under its common name celecoxib. In addition, as of this writing, Pfizer still sells the identical drug under its trade name, Celebrex.

## Dressing Well

Why, you might ask, would they do all that? The answer is that the launch of an authorized generic without using the brand Celebrex but only the generic name celecoxib, is there to capture part of the lower-priced generic market. The reason for continuing sales of the drug under the brand Celebrex is simple and has to do with another aspect of IP law: trademarks.

While Pfizer no longer has its patent, it still controls the trademark Celebrex, which Teva cannot use. Trademark law says that if trademarks continue to be actively used, they, in contrast to patents, do not expire on a date certain. Think about the mouthwash Listerine®. It was invented in the late nineteenth century and given the name Listerine in honor of Joseph Lister, an English physician who had earlier done basic research on microbial infections. We pretty much know what is in Listerine: sodium fluoride plus some essential oils and additional inactive ingredients such as alcohol. It is not patented, and anyone can try and duplicate its original chemical composition. Its century-long dominance in the marketplace is due not to patents but to the valuable name Listerine, a protected trademark of J&J. Competitors can sell identical mouthwashes but never under the name Listerine. They also cannot sell their mouthwash in a uniquely shaped bottle that, by its recognizable silhouette (known in the jargon as its "trade dress"), connotes Listerine. If they do, they will get an immediate letter from the legal department at J&J telling them to cease and desist from using its trademark and its trade dress. This is perfectly fair and legal.

The same is true with Celebrex. With its valuable drug name in hand, Pfizer can keep riding the market it has established over the lifetime of its now expired patent. Of course, Pfizer is no longer alone. There are dozens of generic companies that sell celecoxib, the identical drug, and at much lower prices. The question is: Is this good for patients? The answer is yes, although confusing. At this writing, Pfizer's brand-name Celebrex retails at CVS Pharmacy (without coupon) for $477.20 per thirty capsules, or $15.91/capsule.[5] The same-dosage generic celecoxib retails at CVS Pharmacy (without coupon) for $64.63 per 30 capsules, or $2.15 per capsule.[6] The generic celecoxib is eight times less expensive per capsule than the brand-name Celebrex. That's the generational gift.

The question immediately arises, however: How, with an eightfold price difference, can Celebrex even compete with the generic celecoxib? The answer is, it cannot and does not compete on price—although it does offer coupons worth 80–90 percent discounts. Pfizer's primary strategy for charging an eightfold premium over the identical generic is based on its valuable drug name, its recognizable company name, and its elliptical logo enclosing the word "Pfizer." Only Pfizer can call the drug under the brand "Celebrex," which, of course, is prominently marked with the customary "®" at the upper right-hand corner of the name. And only Pfizer can sell the drug in its recognizable trade dress: a blue-and-white color bottle with the company's valuable logo and name at the lower end.

You might want to ask yourself: Would you pay Pfizer eight times the price of generic celecoxib to get the same drug from Viatris or Teva? I would not. Yet it is worth remembering that Pfizer's strategy is not so different from what goes on in our marketplace for all kinds of commodities. And yes, a drug that has gone off-patent has become a commodity. And as a commodity, it is now in a very different market, where competitive advantage is determined not only by price but also by imponderables like trust and fame. Take shirts, a clearly unpatented commodity. A standard man's white shirt sells on Amazon for about $17.00, yet a similar one by Calvin Klein sells for $75.00, four times more. Of course, drugs are very different commodities from shirts; you can't get sick or die from wearing the wrong shirt. And shirts don't prevent inflammation. But the marketing tactics of Pfizer and Calvin Klein, using whatever IP is available to them, are similar. Pfizer's approach, while more extreme than Calvin Klein's, must work to some extent; otherwise, Pfizer would not keep at it. And to keep at it, Pfizer markets assiduously, trying to convince the public that its famous Celebrex is somehow more trustworthy than the generic celecoxib.

A marketing strategy of not trusting generic drugs may appear dubious. Yet having confidence in all generics without asking questions is also misplaced. Generics are not free of controversies. Farah Stockman, in a well-researched article in the *New York Times* in 2021, wrote about such controversies.[7] Sometimes, after a drug goes off-patent there are so many generics in the marketplace that prices fall too much and generic companies drop out. This, in turn, may lead to drug shortages. At other times the FDA, which has the responsibility of inspecting and approving generic drug manufacturers outside the United States, fails to do so properly, leading to possibly dangerous outcomes. Stockman's investigations showed that some generic drugs made overseas were highly comparable to the name brand, while others had dosages that changed from one batch to the next. There are at this writing close to twenty different companies that sell generic celecoxib, not including Viatris, which sells Pfizer's authorized generic.[8] Quality control, steady supply, and true equivalency among all these companies are not issues to be taken for granted.

The bottom line is that even taking the potential troubles of too many or not carefully regulated generics into account, consumers get the much less expensive and equally effective generic celecoxib at about eight times less than the branded Celebrex of Pfizer. If celecoxib were still on patent, there would be no generics and Pfizer's price for it would be much higher than it is today. The price differential and widespread availability fulfill the promise of Judge Rader's generational gift. Pfizer, riding the wake of a quickly receding time of patent protection, tries the best it can to keep selling its own celecoxib by wearing its blue-and-white trade dress with the elliptical logo on its lapel, the name "Celebrex," expensive marketing campaigns, and some fancy quickstep.

Like two biological adversaries, brand-name companies and their generic foes have evolved strategies to outfox each other. Every time a patent holder evergreens a top drug with a new patent, the generics try to come up with a way to get around it. And every time a company like Teva or Sandoz launches a generic, the brand-name company tries to launch an authorized one. To add even more spice to the mix, it is becoming common to see classical generic companies get patents to protect inventions generated in their *own* research labs. In 2021 there were two court cases that pitted Teva against Eli Lilly, with their roles inverted: Teva had the patents, and Lilly was accused of infringing them.[9] It is amusing to read the full-fledged defenses that Lilly mounted against its arch enemy Teva. Lilly sounded like a generic and Teva like a brand-name company. The two adversaries had known each other for so long that they sang the same tune, except in reverse. In the end, in a Solomonic outcome, Teva won one case and Lilly the other.

The two sides of the traditional pharma world are locked in a never-ending struggle to outwit each other. We, the bewildered public, watch the spectacle from the sidelines. When, at the end of patent lives, the generics enter the marketplace, we should at least understand what is going on when we make our buying decisions: whether we choose to continue buying the expensive branded drug sold by the brand-name company or the more inexpensive generic version sold by a competitor, or the authorized generic sold by the brand-name company itself.

You wouldn't be far off if you concluded that many court battles flare up at the end of patent lives. That is the time when a pharma patent holder has exploited the exclusivity period to its maximum and its profits are the highest. It is also the time when the generic companies lick their lips in expectation of Judge Rader's generational gift. Yet, while waiting, they also dry their powder behind closed doors in anticipation of possible wars. Rader's gift is a nice concept, but sometimes the gift givers are not as altruistic as we would like them to be. At times, a patent at the end of its life feels more like a toy in the hands of a kid who does not want to let go even though his mother tells him that the time has come.

As my law school professors said, a patent is a legally controlled monopoly, limited in scope and time. In my view, it is worthwhile for society to have the Jeffersonian "embarrassment" of a limited and predictable monopoly to promote costly investments in the life sciences. The results of such controlled monopolies are an ever-increasing supply of better drugs for deadly diseases. At least lengthy interferences like the one used to determine who first invented Enbrel, which we discussed in chapter 4, are no longer the law since 2012. In addition, US patent lives based on applications filed after 1995 now expire twenty years from *filing*, not the longer seventeen from *grant*, as was the case with applications filed before 1995. And applications pending at the Patent Office are no longer secret, so the anxiety about submerged submarines has diminished. (As I will show in the next chapter, an occasional biotech U-boat still does breach the surface every now and then, causing major shock waves.)

All these exceptional American legal procedures used to create a great deal of uncertainty as to who would own the patents and how long they would be around before lapsing. For many years, these now archaic procedures threw fog into the limited and predictable monopoly. They are gone, and that is progress. What remain, and what are of less obvious value to patients, are the evergreening maneuvers by the brand-name pharma companies to extend exclusivity. I don't blame them for trying. But when the predictable and limited monopoly becomes less limited and less predictable, our present-day patent system moves away from the benefits envisioned by the Venetians in 1474, legislated by the British Parliament in 1624, and cautiously endorsed by Thomas Jefferson in 1813.

Let's now focus on a set of drugs unique to the biotech age, the biologics.

# 19

# The Age of Biologics

Perhaps the most promising of the many inventions emerging from contemporary biotechnology is the ability to make new drugs by isolating from nature materials such as proteins and improving them by careful genetic engineering. The resulting medicines go by the name of "biologics." Because they are exact or closely re-engineered copies of natural molecules, biologics are powerful medicaments that are exquisitely free of side effects. Let me introduce some biologics, explain their differences from classical pre-biologic drugs, and tell you about a whole slew of new patent questions.

## Assembly Lines

You may remember the anti-inflammatory drug Celebrex that we met in the previous chapter. Both Celebrex and its identical generic celecoxib are drugs known in the jargon as "small molecules." The name reflects their relatively small size compared to biologics. In contrast, biologics, like the antibody fusion drug Enbrel used for rheumatoid arthritis that we met in chapter 4, are about 1,000 times more massive than small molecules like Celebrex. That's roughly like comparing cherries to watermelons. And just as a brand-name small-molecule drug like Pfizer's Celebrex has its "generic" celecoxib, the brand-name biologic drug Enbrel (made by Amgen), has its "biosimilar" Erelzi (made by Sandoz).

The size disparity between the cherry-like pre-biologics and the watermelon-like biologics is not the only thing that makes them different. Small molecules are also known in the jargon as "synthetics." The reason is the technology by which they are made. As their name indicates, such small molecules are synthesized step-by-step by classical methods of organic chemistry. These methods are entirely different technologies from those used to make biologics. Synthetic methods are also more readily available and therefore more easily reproduced to make generics. Once a synthetic drug's formula and its method of preparation are published, most competent organic chemists can replicate them. A pharma company wishing to make the generic copy of a small synthetic drug needs access to relatively few if any secrets or proprietary methods of the original inventor.

Not so with large biologics or their biosimilars. The manufacturing methods for these large molecules involve live cells as sources, multiple fermenters where they grow, and elaborate extractions and purifications to obtain the final products. The methods are best described as "biological assembly lines."

Let's examine, in a very superficial manner, the biological assembly line for making Enbrel. You may recall from chapter 4, figure 4.1, that the protein Enbrel is the artificial fusion of a portion of a human Tumor Necrosis Factor (TNF) receptor linked to a portion of a human IgG antibody. You may also remember that I described the early idea of what would eventually be Enbrel as a "decoy" that circulated around blood vessels soaking up the highly inflammatory molecule TNF. Like a high school proctor monitoring the hallways for troublemakers, the decoy would take TNF out of circulation and prevent inflammation. To make Enbrel, the DNA for the TNF receptor portion is first obtained from a human cell line. The DNA for the human IgG antibody is separately obtained from published data, synthesized, and fused to the portion of DNA encoding the TNF receptor portion. The fused DNA is introduced into isolated Chinese Hamster Ovary (CHO) cells, which then produce the protein during fermentation. A master cell bank made of CHO cells is prepared, stored, and used in a fermentation tank. The sterility, and freedom from contaminants and viral safety, as well as the genetic stability of the cell bank and other production cells need to be constantly verified and controlled. Because Enbrel is produced in cell culture, all raw materials from animal origin used as nutrients need to be carefully controlled so that they do not present a health risk. There needs to be continuous testing for microbial contamination, for toxins, and to maximize the main product while minimizing side products. The processing and purification of Enbrel that end the assembly line is a sequence of separation and ultrafiltration steps, including a critical step of removing viruses.[1]

One consequence of this complexity is that most of the companies that have tried their hand at making biosimilars are not the traditional generic pharma companies of the world but are biotech or large pharma companies themselves, such as Amgen, Biogen, or Sandoz. Another consequence of the complexity is that the international regulations for receiving approval for biosimilars are more intricate than those for small-molecule generics. In fact, biosimilars are not the generic version of biologics; that is, they are not identical to their biologic counterparts. The stability and efficacy of biosimilars depend on their methods of manufacturing. Thus, biosimilars are rarely if ever indistinguishable from the original biologic. Moreover, whether they are "similar enough" to be approved for administration to patients is a tortuous legal proposition.

The number of patents protecting the original biologic has increased manifold over the numbers of patents protecting small-molecule synthetics. Since reproducing the methods of manufacturing biologics is critical to achieve biosimilarity, everything on the assembly lines gets to be patented and potentially litigated: cell cultures, contamination controls, purification methods, the final product—the works. Patent thickets protecting the original molecule and every possible step in its assembly line are now common for biologic drugs. US patent thickets have delayed entry of competitive biosimilars, keeping prices higher than in markets, like Europe, where biosimilars have entered soon after the original patent's expiration.

The story of Enbrel is a dramatic case in point.

## The Enbrel Submarine

You may recall that in the mid-1980s I was a young patent attorney representing Massachusetts General Hospital. One of its scientists had invented a molecule that conceptually would become a precursor of Enbrel—although it turned out not to be Enbrel itself. At about the same time, three additional groups made similar inventions, and there followed a multi-party priority battle to decide who would hold the patent rights. Immunex, of Seattle, Washington, acquired everyone's rights and became the sole holder. Amgen eventually bought Immunex and, through licenses, ended with Immunex's rights for Enbrel. Once Professor Bruce Beutler of the University of Texas was judged to be the first inventor of Enbrel, Texas licensed its patents to Immunex and Immunex sublicensed them to Amgen.

Beutler's basic patent on Enbrel issued in 1995.[2] The FDA approved Enbrel around 1998. The Beutler patent, then sublicensed to Amgen, still had fourteen years to go and expired in 2012, seventeen years after its issuance. That year would mark the end of the controlled patent monopoly; the generational gift of Judge Rader would take effect, and in 2012 a biosimilar of Enbrel would then be available for all to make and use. Right?

Not so fast.

You may remember that before the US patent law changed in 1995, the life of a US patent was seventeen years from the issuance date, not twenty years from filing as it has been since and is today. Critically, any patent applications that were filed before 1995 remained secret; they were the "submarines," which I introduced to you in the previous chapter. Immunex took advantage of submarining and, in stealthy mode, took a license for a pending patent application on Enbrel, owned by Hoffman-LaRoche. That application had been filed before the law changed in 1995 and was therefore still secret. And, to keep matters quiet, Immunex also kept the application under wraps. The Roche application finally issued in November 2011.[3] This date is a few months before Beutler's original patent was to expire in 2012. And because of the seventeen-year lifetime of the new patent, which had been filed under the old pre-1995 regime, it legally extended patent protection for Enbrel until . . . 2028! Like a whale breaching Alaskan waters, and to everyone's surprise, the Roche patent surfaced with a big splash.

I was stunned. Around 1987 when, as a young attorney, I filed Mass

General's patent application, I expected that after the priority litigation ended, the life of the patent of the eventual winner, which turned out to be Professor Beutler's, would end somewhere around 2010 to 2012, that is, twenty-three to twenty-five years later. That would account for six to eight years of litigation plus the seventeen-year life of the winner's patent. I did not expect that patent protection for Enbrel could last until 2028, forty years after my original filing. Yet as Roche's secret U-boat, navigating under the surface for many years, emerged from the depths of the Patent Office, the competition was rudely awakened to the fact that they would have to wait seventeen more years before entering the market with a competitive Enbrel biosimilar. Everyone in the health-care industry dealing with rheumatoid arthritis was taken aback by the realization that prices for this groundbreaking drug were not likely to come down any time soon.[4]

The four decades of patent protection for Enbrel neatly coincided with the years of my professional life as a patent attorney. By the time the Roche submarine surfaced, I was an experienced old hand at this and, however reluctantly, had to allow that the legal departments at Roche, Immunex, and Amgen had pulled off a shocking success for their trio of companies. Immunex's patent attorneys were not only clever in acquiring a license to Roche's submarine application and keeping it submerged, but they also carefully arranged it so that neither Amgen nor Immunex owned the patent itself. Neither of these two companies had outright ownership; all they had were licenses. This subtle legal maneuver made all the difference in the end. The distinction between fully owning the surfacing Roche patent and only licensing it was a central theme of the litigation that ensued after its surprise issuance in 2011.

No one smarted by the appearance of the Roche patent more that Sandoz Inc., a Swiss pharmaceutical company that is known for its manufacture and sale of generics and biosimilars. Sandoz had developed a process for making a biosimilar of Enbrel, which it called Erelzi, and had applied to the FDA to receive approval for its marketing and sale. Sandoz figured that the patent life for Enbrel would end in 2012 and it could then launch the biosimilar. The surfacing of the Roche patent spoiled those carefully laid plans. Immunex, Amgen, and Roche sued Sandoz for patent infringement of the new patent. In defense, Sandoz argued in court that the Roche patent was invalid for what is known in legal jargon as "double patenting," a well-established doctrine of US patent law. The law frowns upon a *single company* owning two patents that are generally to the same invention and that expire at different times. The reason is to prevent an extension of patent exclusivity for inventions that are not distinguishable from each other; that is, that are "obvious" variants of each other. The penalty for the sin of double patenting is that the life of the later-to-expire patent gets cut back to that of the earlier one, so that both patents end up expiring at the same time. Simultaneous expiration means no extension of patent life for obvious variants of the earlier patented drug.

But the prohibition to "double patent" exists only when *one company* owns both patents. If two separate companies own the patents, there is no problem with double patenting. Immunex had arranged the transactions with Roche so that it acquired only an exclusive license to the Roche patent, not ownership. Knowing that the courts are not easily fooled and can tell if a transaction, even if called a "license," is really a sale, the lawyers arranged it so that Roche did remain the owner and not just in name.

Under the agreement, Roche retained certain rights in its patent. For example, the license allowed Roche to do research with Enbrel, or gave it the ability to start patent litigation against infringers if Immunex and/or Amgen declined to do so. Such reservation of rights is inconsistent with Immunex having full ownership to the patent. It is akin to me, as owner of a house, "selling" it to you but allowing me to come in and spend weekends there while you are out of town. And if you do not promptly expel any uninvited house squatters, I can come in and do so. Under such circumstances, a judge would not likely conclude that the transaction that we called a "sale" was indeed that. A judge would conclude that you did not own the house outright: it would be clear that no matter what we called it, I continued owning it and you only had a lease, subject to my retained rights.

That is what happened in the Enbrel patent litigation. The Court of Appeals dissected the transaction between Immunex and Roche and in 2020 concluded that the Roche patent was still owned by Roche, and not by Immunex or Amgen. Sandoz's double patenting defense, which was based on common ownership of the Roche patent and some earlier Immunex-owned patents on Enbrel, failed.[5] As a consequence, the Roche patent was held to be valid and to protect Enbrel until 2028.

I was frustrated by the unavailability of the expected generational gift of Enbrel's biosimilar Erelzi. Yet I reluctantly had to admit to the able legal maneuvering by the Immunex, Amgen, and Roche legal departments. I am an attorney with a social conscience, but even those of us with conscience may secretly admire intelligent lawyering when we see it. So, while my social conscience was taken aback, my trained legal mind grudgingly accepted the sudden appearance of the Roche U-boat and its survival at the Court of Appeals.

Is all this crafty maneuvering legal in the United States? Yes, perfectly legal. How about in Europe? Not so much. Over there, most of these things don't work, because the law is and has been different. There have never been any submarine patent applications pending in the European Patent Office; everything has always been transparent and competitors can see what is in the files of the office in Munich. Patent life has never been seventeen years from issuance but twenty years from filing. Because the life of a European patent starts from its day of filing, it has never been in anyone's interest to keep European applications pending for years and years. Long pendencies eat up patent life. Professor Beutler's original patent for Enbrel, filed in 1987, expired in Europe in 2007. And the European Roche patent filed in 1990 expired in 2010. Biosimilars to Enbrel started appearing in Europe a few years later and, according to commentators, have gained about 40 percent of the Enbrel market, significantly bringing down drug prices.[6]

After the US law finally changed in 1995, our patent system is now much more like the European one. However, because of the differences in legal regimes still in existence during the late 1980s, when all the fundamental Enbrel patent applications were filed, the late-surfacing Roche submarine has kept US patent protection for the biologic going for almost two decades longer than across the Atlantic. So, my hat comes off for the

Immunex, Amgen, and Roche US patent lawyers, who have been so artful in their use of our now outdated submarining habits. My conscience, however, remains perturbed by the financial consequences of their maneuvering to US patients in need of Enbrel.

## The Humira Thicket

In an ironic twist on the Enbrel patent story, Amgen found itself "on the other side of the v." when, in 2016, it confronted 132 patents protecting Humira, a monoclonal antibody and its manufacture and uses that were patented, made, and sold by AbbVie, of Chicago. Amgen wished to come into the market with a biosimilar to Humira, which it named Amjevita®. Amgen knew that the basic Humira patent would expire in 2016 and, like Sandoz, which hoped to launch a biosimilar to Enbrel when Professor Beutler's basic patent on Enbrel expired in 2012, Amgen hoped to launch Amjevita in 2016. And, while Sandoz's biosimilar Erelzi confronted a submarine patent owned by Roche, controlled by Amgen, and that blocked Erelzi's US entry until 2028, Amgen confronted a large bundle of AbbVie patents on Humira—a thicket as it were—the last of which would expire in 2034.

This time the shoe was on the other foot. Amgen did not hold the patent cards; those were held by AbbVie, and AbbVie promptly started patent infringement litigation against Amgen and several other biosimilar company hopefuls. The litigations eventually settled, and all aspirants agreed to delay their biosimilars' entry until 2023, a seven-year delay in launch. In turn, AbbVie gave up all enforcement of its patent coverage beyond 2023, forfeiting a possible eleven years of remaining patent protection.

Things got really interesting when in 2020, the mayor and City Council of Baltimore, which administer welfare-benefit plans that pay for Humira, sued AbbVie in Illinois, arguing that the very existence of the 132-patent thicket protecting the biologic was an illegal attempt by AbbVie to monopolize commerce in the drug. It was in violation of the Sherman Antitrust Act, they said. If they succeeded in convincing the court that the very existence of the thicket was illegal, they hoped to be reimbursed by AbbVie for what they called the "monopoly prices" they had incurred in paying for welfare benefits. They did not argue that each of the 132 patents was invalid, or that they had been obtained by fraud on the Patent Office and thrust on the city. They argued that 132 patents are "just too many for anyone to hold," as Judge Easterbrook of the Seventh Circuit Court of Appeals curtly said in his decision denying the Baltimore group the relief it wanted.[7]

The judge was almost contemptuous of the city of Baltimore's theory. "What's wrong with having lots of patents?" he asked. "If AbbVie made 132 inventions, why can't it hold 132 patents?" And the judge further held that the settlement between AbbVie and Amgen had not been a conspiracy to hurt the public. He took pains to distinguish situations in which the delay in entry of a generic company is induced by the patent owner when it pays the generic to forego the market for a few years. This is known in the jargon as a "pay for delay" maneuver. In such a situation, the critical legal question becomes: How much did the pharma company holding the patents pay the generic company to defer entry? When the amounts changing hands are seen to be a portion of the spoils from a market-splitting arrangement between them, the settlement might be anticompetitive and violate the antitrust laws.[8]

But that is not what happened here, said the judge. AbbVie did not pay Amgen or anyone else to delay entry. The judge was not convinced that AbbVie had left money on the table in exchange for prolonging its patent coverage to prevent competition. The judge held that this had been a classic quid pro quo entered to settle complex litigation. Every party gave up something, and the case ended. Amgen's Amjevita finally launched in early 2023 and is now providing price competition to AbbVie's Humira.

## Pharmacy to the World

The Humira case shows how hard it is to argue that there is something illegal in clustering multiple patents to protect a blockbuster biologic. If a patent is filed on every step of a biologic's assembly line, a solely owned thicket may well develop; yet such thickets are not unlawful per se. Unless there is some sort of "pay-for-delay," or one or more of the patents in the thicket can be shown to have been procured by fraud on the Patent Office, for example, by lying about basic facts, the chances of victory in court are slim to none. The US patent code does not prohibit thickets either. It does prohibit the "double patenting" of similar inventions that we discussed earlier in the Enbrel case, and it is meant to assure that later filed applications are only for "non-obvious" improvements over the earlier drug, but as we have seen, clever lawyering goes a long way to getting over such hurdles.

Other nations' patent codes do limit the amounts of patents that a company can use to protect a single drug. The best known one is India's patent law, which says that new uses for, or modifications of, known drugs are patentable only if "they differ significantly in properties with regard to efficacy."[9]

The Indian law was severely challenged in 2013 by the Swiss company Novartis in the case of Gleevec®, one of the most successful small-molecule anticancer drugs ever. Gleevec is not a biologic, but the lessons from this court decision will apply in any future patent ligation in India. In the late 1990s Novartis patented Gleevec everywhere, except India. It could not do that, because at the time Indian law prohibited patenting pharmaceuticals, period. When that law eventually expired, Novartis tried to patent a variation of the basic Gleevec drug, but by then an earlier form of Gleevec was in the public domain. Citing the words of the law, the Indian Supreme Court ruled that the later substance that Novartis sought to patent did not differ "significantly in properties with regard to efficacy" from the earlier one. It did not matter to the court that Novartis could not obtain an original patent because the law in India at the time prohibited it from doing so. The court insisted that Novartis could only receive a single patent on Gleevec, and since it did not receive one, it could not patent a modification without demonstrating better efficacy.[10] It is clear that India, which is known as a "pharmacy to the world," showed that it will go to great lengths to protect its booming generic, and (I presume) its budding biosimilars, industry.

The Gleevec judgement was strongly supported by Doctors Without Borders and the World Health Organization, both of whom publicly oppose the evergreening of pharmaceutical patents. In contrast, the judgment was strongly opposed by the Pharmaceutical Research and Manufacturers of America (PhRMA), a trade group that supports strong IP protection as critical to the invention of new drugs. PhRMA will lobby energetically against anything, anywhere, that would shorten patent lives or inhibit maneuvers such as evergreening.

What is also clear to me, however, is that the lack of submarine patents, of large solely owned thickets, and of less instances of evergreening do not seem to have inhibited pharmaceutical research in places like Europe. Pharma companies with strong European research presences—such as Novartis, AstraZeneca or Roche—have continued inventing and giving us lifesaving medicines. Innovation in Europe does not seem to have dried up, as PhRMA warns.

Any attempt to legislate something like India's strong anti-evergreening law in the United States will likely fail and, in my view, should fail. India may well be the pharmacy to the world, but the United States remains the source of many of the most successful biologics on earth—a pipeline that we do not want to turn off.

The problem is not so much evergreening as maintaining strict principles of "non-obviousness" for follow-on pharmaceutical inventions. The Patent Office and the courts need to remain vigilant so that our patenting standards for what is a "non-obvious" drug improvement are applied strictly. Many new formulations, new delivery modes, or modifications of preexisting drugs are not as inventive as resourceful patent lawyers propose them to be. They are not deserving of the additional patents that, if issued, will create dense thickets. And, keeping in mind the lessons from the Humira litigation, our courts also need to be alert to any attempts to enforce legal clusters of patents in an illegal manner, such as in pay-for-delay schemes.

The tales of Enbrel and Humira give us a perfect introduction into the patent landscape for biological drugs in the developing world. Among these drugs, biological vaccines such as the mRNA anti-COVID-19 ones are of particular interest. We get to this next.

# 20

# International Access to Patented Biologics

Let me forewarn you that the differences in perception of the patent system between those in the Global North and those in the Global South are, to put it mildly, fierce. The dissenting voices can be heard across continents. (My use of the North-South nomenclature is not strictly a geographical one, but it will serve as a shorthand for "richer" and "poorer.")

I experienced this firsthand one day as I was lecturing on technology transfer policies to a group of German pharmaceutical lawyers in Frankfurt. Toward the end of my lecture, I pronounced the two dirtiest words ever afflicting the ears of executives from Big Pharma: "compulsory license." I said in passing that if the advanced drugs they patented did not make it into Latin America, the governments there might force a compulsory license. You could hear a pin drop after I said it. My host looked at me from the auditorium and, in a stress-enhanced German accent, said icily, "We don't discuss compulsory licenses." And that was that.

## Compulsory Licenses

A compulsory license is the right of a government to compel a private maker of a patented drug, against its wishes, to allow another company to make the drug at a much lower price. The "march-in" and "taking" provisions available to the US government in cases of underuse of, or critical need for, a patented invention that I discussed in chapter 18 also result in compulsory licenses. In that chapter they were in the United States: here they are worldwide. The fear and loathing of compulsory licenses are not limited to Big Pharma from the developed world. They also exist on the other side of the great pharmaceutical chasm between the Global North and the Global South.

Many public health officials, government ministers, and academic advisors in the Global South have embraced compulsory licenses as the way to get access to patented medicines. Consequently, the ideological and legal clashes about such licenses between northern pharmaceutical interests and southern government policies are intense. And even when the governments of the Global South embrace compulsory licenses, they are not entirely free from political anxiety. I have witnessed international pharma companies recruit a local embassy to push for more stringent enforcement of patent rights and less talk of compulsory licenses. Similarly, when the embassy is American, the disparity of power is so large that most small countries of the world drop the notion of compulsory licenses altogether.

I am a believer that because of patents and a strong judicial system to enforce them, the United States is the country of biological and pharmaceutical innovation par excellence. Very few come close to us in generating, through invention and investments, not just biological drugs but biotech *companies*. I know of countries with excellent chemists and biologists but poor patent or judicial systems that have innovated very little. They harbor local pharmaceutical interests that, wittingly or not, foster economic dependence. Mostly, they copy technology from the countries of the Global North, and, with the help of friendly legislators, undermine efforts to strengthen their own patent systems.

"If the biotech companies of the US and Europe invent, let them get their patents over there," the argument goes. "But let's not give them strong patent rights here. Drug prices will be high, and lifesaving new therapies will not be readily available to us. If we keep local patent rights weak, we will make and sell the drugs cheaper than they do. And that will benefit the local population." For the local pharma companies, free copying is much more profitable than investing in innovation. If they can make a state-of-the-art drug without paying anyone a license fee and take credit for helping the local population's health needs, they are local heroes. The result of this misdirected policy is continued reliance on the innovators of the North to provide the latest pharmaceutical inventions. And that is not a road to economic independence.

The next argument, especially of countries in the Global South that do have some semblance of a patent system, however, is more difficult to rebut. It goes to the core of not having ready access to patented medicines. The ultimate bad scenario in this picture is northern Big Pharma getting local patents. Let's say that a breakthrough therapy for cancer has been invented by an English biotech company and patented worldwide, including in Malaysia. Yet, because the English company considers the Malaysian market too small, the English don't manufacture it there, and if they import it from England, they sell it at very high prices. Plus, the existence of their Malaysian patent blocks importation from an Indian generic company that would be willing to make the drug and bring it in. The Indian company doesn't want to be sued in Malaysian courts for patent infringement by the English patent holder. The local patent functions essentially as a private import license in the hands of the English company. Consequently, the latest cancer drugs in Malaysia are too expensive, or worse, not even available.

The rich of Kuala Lumpur can afford the drugs or get on a plane to London and get treated there. But what about the Malaysians who can't afford costly therapeutic outlays? What are Malaysian public health officials to do? This is where international compulsory licenses come in, through an agreement called TRIPS.

## Human Rights

TRIPS, "The Agreement on Trade-Related Aspects of Intellectual Property Rights," became effective in 1995. Pretty much every country in the world signed it. It requires every signatory to enact patent systems along the lines of those of the United States and Europe: strong protection, including for pharmaceuticals. Everyone must respect the private IP rights of their own nationals, as well as those of foreign holders. The 1995 TRIPS agreement was a sort of one-size-fits-all, from Austria to Zimbabwe, regardless of development, poverty levels, or endemic health conditions. At a global level, it reflected the developed world's view that "patents are a good thing." There was not much in the 1995 TRIPS agreement in terms of facilitating access to patented drugs.[1]

But then, the AIDS epidemic shook things up. Because of AIDS, by the mid-2000s the life expectancy at birth in many African countries had dropped by an average of about ten years, from 60–65 years of age in 1985, to 45–55 years in 2005. This was a health crisis of earth-shaking magnitude.[2] Meanwhile, the big pharma companies of the North were inventing and selling lifesaving anti-HIV/AIDS drugs. Their availability in Africa, especially in South Africa, however, was low; prices were high, and patents stood in the way of local manufacturing or generic importation.

In 1997, in response to this growing health crisis and under the broad outlines of TRIPS, South Africa enacted a law, the "Medicines Act." This law allowed compulsory licenses for critical drugs so that they could be imported from less expensive suppliers. In other words, whether the pharma companies liked it or not, they were compelled to let other companies import and sell the drugs in South Africa. In 1998 forty big pharma companies, unhappy with the law, sued the government of South Africa arguing that, among other things, the Medicines Act violated TRIPS.[3] This lawsuit was perhaps the biggest blunder Big Pharma has ever made. I wonder whose misguided advice they took when they decided to start the South African litigation. It backfired so dramatically that they are today living with the commercial and ideological consequences.

To make an extraordinarily complex story simple, the lawsuit, as never before, mobilized the developing world against the pharmaceutical industry. An activist group called Treatment Action Campaign (TAC) entered the litigation as a friend of the court on behalf of South Africa. Instead of arguing the ins and outs of the TRIPS agreement, which was comfortable territory for the pharma plaintiffs, TAC shifted the argument to be one of human rights: the right to health, it told the court, trumps corporate IP rights. TAC contended that the fundamental dignity and life interests of South African citizens were at stake.[4] It was a brilliant legal and strategic maneuver, and it changed the conversation forever.

There was astonishing international support for TAC: days of action; demonstrations in multiple cities around the world; 250 organizations from thirty countries who signed a petition opposing the lawsuit; Doctors Without Borders collecting a quarter of a million signatures against the suit; the European Union urging the suit to be dropped; and Nelson Mandela, by then the former president of South Africa, openly criticizing the high prices of AIDS drugs.[5] This global mobilization and censure shamed the big pharma companies to drop the litigation in 2001. No pharmaceutical company wants to be accused of violating human rights, especially not one that hopes to sell in multinational markets. Transmuting a legalistic discussion on the boundaries of the TRIPS agreement into a struggle for human rights changed the discourse.

Inspired by this turn of events, and feeling the wind of public opinion behind them, in the early 2000s, multiple signatories to TRIPS issued what is known as the "Doha Declaration." Doha would give each signatory country the rights, during a public health crisis, to force compulsory licenses for critical patented medicines. Doha became part of an amended TRIPS agreement, which was finally ratified by two-thirds of the parties in 2017. Complying with TRIPS and its compulsory licensing of critical drugs is now the obligation of every signatory country. If you suspect that the developed countries with strong pharma lobbies resisted the 2017 Doha amendment each step of the way for close to fifteen years, you would not be mistaken.

## Flexibilities and Complexities

The updated TRIPS agreement is a careful balancing act. It is like an Alexander Calder mobile that has come to rest, at least for the time being. The developing nations may get access to patented drugs through compulsory licensing. These provisions, in the jargon that has since developed, are known as "TRIPS flexibilities." Sure, a developing country must respect patent rights, but it has *flexibility* in getting around them.

In return, the developed world negotiated hard for things that it favored. The best example of this is the roadmap to compulsion. TRIPS has an agreed-upon outline for what must happen before a compulsory license can be issued, including some vague provisions. In cloudy legalese the provisions say that a compulsory license may only be issued if the requester has made prior efforts to obtain authorization from the patent holder on "reasonable commercial terms and conditions" and that such efforts have not been successful "within a reasonable period of time." And they add that the patent holder must be paid "adequate remuneration" for the compulsory license.[6]

All this nebulous terminology—"reasonable terms and conditions," "reasonable periods of time," and "adequate remuneration"—comprises concepts that might grip the imaginations of international IP lawyers. For the rest of the world, however, they sound like, and are, a recipe for haggling. Not surprisingly, they are known in the jargon as "TRIPS complexities." Sure, the pharma companies may be compelled to grant licenses they do not wish to grant, but they can set up *complex* roadblocks that eat up time, effort, and money.

The agreement tries to juggle the TRIPS "flexibilities" with the TRIPS "complexities." It's not an easy job to provide access to patented medicines in the Global South and not undermine pharmaceutical innovation in the Global North. The basic argument of the pharmaceutical industry has always been that expanding compulsory licenses to lower drug prices would negatively impact investment in research and development of new medicines.[7] Frankly, I am never sure how well grounded in actual numbers are the threats to Big Pharma that compelling licenses in the Global South will harm their R&D pipelines in the Global North. The intricacies of cost accounting in Big Pharma are opaque at best and, even if readily accessible, are beyond my limited powers of economic analysis. I can't be certain if the consequences of compulsory licensing will be as dire as predicted by Big Pharma, but I doubt that they would be cataclysmic.

At this point, it is fair to ask: How has the TRIPS Agreement worked out in practice?

"Better than commonly assumed," answered a survey published in 2018 by the World Health Organization (WHO).[8] The survey identified close to 200 instances of possible use of "TRIPS flexibilities" in the period 2001–16. They covered products for treating fourteen different diseases. Close to 80 percent of the drugs concerned medicines for AIDS, and 20 percent were for other diseases. The authors concluded that use of the TRIPS agreement is an increasingly important way for accessing lower-cost generic equivalents in the developing world.

The attempt at aligning the incentives of a strong patent system with providing fair access to medicines have proven to be a delicate balancing act. Not even the Flying Karamazovs, the famous troupe from California, could easily juggle the conflicting IP interests of pharma companies in the Global North with the health needs in the Global South.

## Patents and the Pandemic

In 2020, in a close replay of the AIDS events from four decades earlier, the COVID-19 pandemic shook up the delicate balancing act achieved by the TRIPS Agreement. The question of whether compulsory licenses should be used by the Global South countries to gain access to patented vaccines became the centerpiece of a debate on health inequalities between the South and the North.

Mark Twain is alleged to have said, "History doesn't repeat itself, but it often rhymes." The COVID-19 crisis of 2020 was not identical to the AIDS one of the 1980s, but it rhymed. One difference is that forty years after AIDS first appeared on the international scene, the world had some experience with TRIPS. A second and critical difference is that the most advanced vaccines of the COVID era, the so-called mRNA vaccines, are not like the antiviral drugs of the AIDS era; mRNA vaccines belong to the superb new class of biologic drugs that we discussed in chapter 19.

On October 16, 2020, India and South Africa, invoking TRIPS, presented to the World Trade Organization (WTO) a proposal for members to force a waiver of IP rights on COVID-19 vaccines.[9] The argument was that the developed countries should bypass the TRIPS "complexities" and go straight to accepting compulsory licenses. The countries of the Global North rejected the proposal outright with the usual arguments that a waiver would de-incentivize research into additional prevention and therapies. Instead, they proposed a program of voluntary transfer of finished vaccines to countries in need.

After the much-publicized October 2020 request, the heated international debate of the COVID-19 pandemic quickly became to waive or not to waive patents.[10] Yet, as I have shown elsewhere, this was a false debate.[11] In their 2020 request to the WTO, the governments of India and South Africa asked for a waiver of all IP rights, not just patent rights. Mustaqeem De Gama, the leader of the South African Permanent Mission to the WTO, said as much (the italics are mine): "What this waiver proposal does is it opens space for further collaboration, *for the transfer of technology* and for more producers to come in to ensure that we have scalability in a much shorter period of time."[12] Note that De Gama sweetened the message about technology transfer by couching it in the language of "collaboration."

The October 2020 request went one step further than the classic waiver of TRIPS "complexities." It went to transferring the still-confidential information of vaccine manufacturers such as Pfizer, BioNTech, or Moderna to companies in India and South Africa. It asked for transfer of technology that was still secret. If there is one thing that biotech companies, large or small, will not readily share it is their trade secrets. Such secrets are more guarded than patent rights. The AIDS epidemic of the 1980s led to an expanded TRIPS agreement that promoted the compulsory licenses of patented drugs by countries in need. What it never did is encourage the wholesale sharing of trade secrets. I am not surprised that the pharma industry did not agree to collaborate voluntarily and share its secret vaccine know-how and technology.

When there are few if any confidences, innovator companies are not adverse during pandemics to grant generic companies licenses for life-saving medicaments. In late 2021 Merck announced a small molecule synthetic anti-COVID-19 drug called molnupiravir.[13] The Merck drug appeared to be effective in reducing hospitalizations of infected patients. The drug's formula and its relatively straightforward method of preparation were published and could easily be replicated. There were few secrets that had to be transferred to generic companies. The simplicity of its manufacturing explains why Merck promptly agreed to license its IP rights in molnupiravir to generic companies.

Yet, as we saw in chapter 19, making a biosimilar of an ultramodern biologic mRNA vaccine was, in 2020, and remains much more complicated than making Merck's synthetic molnupiravir. We saw that methods for making biologics and their biosimilars involve live cells, fermenters, and lengthy and complex extractions and purifications; these are the notorious "biological assembly lines." The biological production line for making Pfizer's Cominarty®, the mRNA biologic vaccine against COVID-19, is complicated much like the assembly line we described for Amgen's manufacture of Enbrel. Making Cominarty is an intricate exercise in biotech manufacturing, followed by worldwide production and distribution.[14]

This complexity implies that even if waivers of secret IP are granted, they would not achieve the immediate need for worldwide vaccination. What is required in waiving IP rights on biologics like the latest mRNA vaccines is much more than patents. It involves technology transfer from the developers of the vaccines to generic companies capable of making biosimilars. Yet even if transfer of such technology happened, approval-worthy biosimilars of the most advanced vaccines would not show up any time soon.

## Easy-Breezy Vaccines

It would be simple if Pfizer or Moderna gave companies in the Global South all the technology they need to ramp up and produce copies of their complicated preparations. But I do not think that will happen in a short period of time. There is, however, a workaround to giving the entire world ready access to effective biological COVID vaccines: make the vaccines less advanced.

In late 2021 Peter Hotez, University professor of biology at Baylor University, Texas, and his collaborator Dr. Maria Elena Bottazzi announced the availability of Corbevax®, a COVID-19 vaccine that while less advanced than the state-of-the-art mRNA vaccines of Pfizer or Moderna, seemed to be equally effective.[15] Corbevax is based on technology one step below the complicated mixtures of mRNA and lipids much desired for and used in the Global North.

Hotez's vaccine is a so-called subunit vaccine. Subunit vaccines are ones that are generated by separating and using selected proteins from infectious viruses. Subunits have a mixed history. You may recall from chapter 14 that I discussed GlaxoSmithKline's subunit malaria vaccine RTS,S, which, even though not as effective as desired, was recommended for mass inoculation by the WHO in 2021. You may also remember that Dr. Stephen Hoffman, disappointed with the effectiveness of subunit vaccines against malaria, turned to his aseptic mosquitoes instead. Yet subunit vaccines against other infectious agents, such as hepatitis B, have been successful. Since its approval in 1986 by the FDA, the hepatitis B vaccine is manufactured in recombinant yeast that produces a subunit of the hepatitis B virus. The subunit gives rise to good immunity when injected. Using the well-understood hepatitis B model, Hotez generated subunit vaccines against COVID-19 in recombinant yeasts. According to press announcements, Hotez's vaccines provided long-lasting immunity against the coronavirus in clinical trials in India.[16]

Most important for our story is that Hotez and Bottazzi announced that they would not seek patents, that Corbevax would be patent free. They believed that low-income countries would be able to produce their subunit vaccines in an affordable and safe manner. Hotez called it a "a low-cost, durable, easy-breezy vaccine that can vaccinate the whole world."[17] By late 2021 the government of India had ordered 300 million doses and plans were underway to make it available in Indonesia, Bangladesh, and Botswana.

Hotez's strategy was not to go for the latest high-tech mRNA vaccines, the ones that were difficult to manufacture without access to the guarded secrets of the pharma world. Instead, Hotez worked to develop (hopefully) effective vaccines with somewhat less than the highest technology. While this sounds like shortchanging the developing world from the most recent advances, it is a good example of designing around patents. If you are blocked by whatever reason, whether by a strong patent or a complicated and secret assembly line, and you encounter less-than-ideal cooperation from the holders of secrets, you work around the obstacles the best you can. Much as we would like it to be the case, effective vaccinations do not necessarily have to be done with the latest mRNA vaccines of Pfizer or Moderna.

There is no wisdom that says that everybody in the world always needs to have the latest innovation. The world needs something that works and that can be made without much effort. Hotez's subunit vaccine technology was purposely left unpatented. It could be made available in a hurry, and if it was equally effective, the patients in the Global South would still get the antivirus protection they needed and deserved. The strategy of not necessarily climbing the highest mountain in the range of available technologies is a variation of what economists call "intermediate technologies." I will have more to say about intermediate technologies in chapter 23.

The urgency of vaccinating as much of the world in as short a time as possible transcends the COVID-19 pandemic. There are dire predictions that societies everywhere need to be ready to deal with future pandemics coming our way.[18] In such situations, what will be needed is an immediate humanitarian response by the developed world followed by carefully negotiated technology transfer agreements and patent waivers.

If (and when) the next pandemic hits, the vaccination efforts ought to proceed in three phases. At the outset, the companies of the Global North that can manufacture vaccines should expand their production facilities and ship vaccines to the developing world. No immediate wholesale tech transfer is necessary, other than to share basic know-how on cold storage, the technology necessary to make sure that the biologic is still active when it gets injected. Not all vaccines require cold storage; apparently, Hotez's subunit one does not. In any event, cold storage technology is not as tightly kept a secret as the manufacturing steps involved in the biological assembly lines. Alternatively, work can also get going on the Hotez strategy of seeking effective vaccines that are not as complex as the latest ones made by difficult-to-reproduce and secret assembly lines.

In parallel, but with less rush, the companies in the Global North should negotiate access to the complex secrets involved in setting up their biological production lines. In such a less urgent manner, biosimilar companies in India and South Africa might be able to negotiate confidentiality agreements with Big Pharma and obtain the know-how necessary to produce approval-worthy state-of-the-art biosimilars. Once that happens, countries in the Global South can become self-sufficient.

And finally, any existing patent rights should be waived, if need be, so that once the biosimilar companies in the developing world are manufacturing, they will not be stopped by patent threats.

With future pandemics possibly around the corner, all of us must come to grips with increased international collaboration. Compulsory licenses for critical pharmaceuticals in the Global South, whether synthetics or biologics, are the new realities in which the big pharma companies are living, whether their executives wish to talk about such licenses or not. These are the unintended consequences of the industry's misguided South African lawsuit of 1998, which brought human rights into the equation.

## Humanitarian Contracts

One more thing on vaccines. In chapter 14, after I shared the tale of aseptic mosquitoes and their use for developing sporozoite vaccines for malaria, I promised that I would get back to the theme of whether patenting vaccines is a good idea. The question is: Should society grant exclusive rights to critically needed vaccines during a global emergency like a pandemic? The dissenting voices we heard in the chapters in part IV may give us some of the answers. The question of whether yes patents or no patents in vaccines need not have a binary resolution. There is plenty of middle ground.

For starters, vaccines are victims of their own success. The more effective a vaccine, the sooner it is no longer needed.[19] Therefore, independently of patent incentives, the pharma industry has not invested in inventing new vaccines as heavily as they have invested in inventing daily drugs. Strong patent protection for vaccines does not always incentivize more investment.

Sometimes, however, heavy investments are called for, especially when the vaccines are advanced biologics. Such is the situation in the development of Sanaria's sporozoite vaccine for malaria or BioNTech/Pfizer's mRNA vaccine for COVID-19. Incentivizing such investments will be harder if society prohibits patents on vaccines. As we have seen throughout this book, holding a patent that is infringed by another company inevitably leads to conflict, which, if unresolved, ultimately leads to patent litigation. However, patent litigation need not prevent access to modern biologic vaccines in the Global South.

A case in point is the lawsuit filed in 2022 by Moderna against Pfizer, both competing manufacturers of mRNA vaccines.[20] Moderna alleged that Pfizer's vaccine infringes Moderna's patents. It added that it will enforce its patents on mRNA in the developed world and that the lawsuit is not against a generic company in the Global South trying to make a biosimilar to its vaccine. In response, Pfizer and its partner BioNTech filed a lengthy defense and countersuit against Moderna, attacking Moderna's patents as invalid. Time will tell what happens in this battle of northern biotech giants. The suit shows that the COVID-19 pandemic has not stopped one company in the Global North from zealously protecting its patent rights against another company in the Global North. Such litigations are what we might properly call "first-world problems." Hopefully, the lawsuit will not affect the supply of vaccines to the Global South. The two issues should be kept apart.

In my view, the default on vaccine patents ought to be to continue granting them yet take advantage of legal and commercial mechanisms that have been created for access of patented products in the Global South. These mechanisms have grown out of a wish to balance companies' IP rights with human rights. They include technology transfer with so-called humanitarian clauses to be included in pharmaceutical and biotech contracts between the patent owner and its license holder. These clauses reserve the right of the owner, often the source of basic technology, to ensure that the licensed invention is made available to Global South countries at lower prices than those that the recipient may charge to customers in the North.[21] Several big pharma and agribiotech entities—such as AstraZeneca, Bristol Myers Squibb, Syngenta, and even Monsanto (while it existed)—have shown a willingness to segment their markets and facilitate access to their products in poor countries.[22]

The market segmentation caused by humanitarian clauses would work well for vaccine patents. In fact, something along the lines of tiered pricing has already developed for vaccines: wealthy travelers from the Global North pay higher prices than are paid for the same vaccines by the local populations of the countries that the Northerners visit.[23]

I have learned that things are never either/or in patent matters. Legal creativity and goodwill can keep the system in fair balance. Pharma companies can get patents with the understanding that human rights—whether compelled by the TRIPS Agreement or by voluntarily segmenting world markets—must be permitted to coexist with IP rights. Patents on vaccines can continue being granted if the owners recognize not only their responsibilities toward shareholders but to the world at large.

# 21

# Genetically Modified Crops

The last dissenting voice we will hear in this part has to do with genetically modified crops, GM crops. We cannot talk about the effects of GM crops without going back to sixteenth-century Perú. When the Spanish conquistadores discovered potatoes there, they were quick to take them back home by the shipload. The potatoes took over Europe, becoming the main source of nourishment in poorer countries like Ireland. When the conquistadores discovered that the Peruvian natives collected guano from islands off the coast and used it as fertilizer, they took that too.

It turns out that a microscopic stowaway in the guano ships, a fungus by the name of *Phytophthora infestans*, hitched a ride from Perú to Europe, with a stop in North America.[1] Because the Irish Lumper, the potato developed and grown in Ireland, was derived from a single variety of Peruvian potato, it was susceptible to the nasty fungus and was promptly wiped out. The blight led to the Irish famine of the early 1850s, which killed close to a million people.[2] Had the farmed potatoes been more biodiverse, other varieties could have survived. The Irish catastrophe provides a major agricultural lesson for contemporary societies: dependency on a single seed variety is dangerous.

## Unintended Consequences

As is the case with Big Pharma, modern agricultural biotech, agribiotech for short, has come under fire for "monopolistic" practices. Some of the consequences of the controlled and limited patent monopoly in the big pharma world, such as temporarily high drug prices, are foreseeable and cautiously acceptable. However, when we talk about agriculture, as we do in this chapter, the consequences of patents, such as the takeover of the world's crops by continuous monocultures, also known as "monocrops," have not been as foreseeable and have turned out to be unintended. They may be downright Faustian. The concerns with the ascendancy of patented monocrops echo the way that the Lumper potato monopolized Irish agriculture and led to calamity.

In chapter 15, I discussed the breakthrough discoveries made in the 1980s in plant genetics, which caused the re-engineering of common crops such as corn or soybeans. These are the greatly admired—or, depending on your point of view, maligned—genetically modified organisms (GMOs). Unlike the pharma or biopharma industries, the agribiotech companies are not critiqued as much due to high prices (although they do get their share of that) but because some of the patented GM crops, specifically those that are herbicide resistant, have indirectly led to economic, biological, and health concerns.

I should start by saying that I have nothing against genetically modified organisms as such. There is no scientific evidence I have seen that has convinced me that GM crops will harm me or cause me health problems. I don't mind eating genetically modified soybeans or eating pork that has been fed such beans. They won't make me sick. The genes that confer herbicide resistance in Monsanto's Roundup-Ready soybeans will not jump into my genome and make me herbicide resistant.

The main issues with herbicide-resistant GM crops are not directly related to the GMOs themselves. They have to do with the indirect and unintended consequences of their use and misuse. I cannot talk about GM crops without involving Monsanto, which for better or worse has been at the center of most controversies. I bring back Monsanto to provide a view of patents from two sides: the acceptable and the Faustian. In the previous chapters we saw how the US patent system could use further improvements to make it more predictable and limited. Yet even with its flaws, and even at the expense of temporarily high prices of things like patented drugs, the system creates a robust environment where inventions flourish. However, the Faustian side of the system reflects the monopolistic consequences of widespread use of Monsanto's patented herbicide-resistant GM crops.

You may also remember from chapter 15 that the invention of herbicide-resistant, genetically engineered soybeans by Agracetus led Monsanto to buy the company in 1996. Monsanto started commercializing Roundup-Ready soybeans the same year. The appearance in the market of Roundup-Ready seeds led to increased sales of Roundup, a Monsanto weed killer that was losing commercial favor because of its lack of selectivity and evidence of ill-health effects in humans who used it. The development and sales of Roundup-Ready soybeans made Roundup selective and revived its flagging fortunes. It led Monsanto to become the agribiotech behemoth we all grew to know and love (or not).

While Monsanto's patent on Roundup, the herbicide, expired in 2000, the patent on Roundup-Ready soybeans did not expire until 2015. Sure, after 2000, when Roundup became widely available, farmers could buy its generic form, glyphosate, from anyone and at lower prices. But the patent on Roundup-Ready soybeans extended and strengthened Monsanto's dominance of the field of Roundup-Ready crops for another fifteen years. It was a smart commercial move for Monsanto, but it had unintended consequences for the rest of us.

Farmers learned to love Roundup-Ready soybeans and other such crops. The weed killer could be applied after crops came out of the soil, thus avoiding the need to till the land ahead of planting. The yields per acre went up, and so did the farmers' profits. Farmers all over the world started planting less diverse crops than before, focusing on the Roundup-Ready crops instead. In Argentina, which, together with the United States and Brazil, is one of the top three producers of GM crops, the production of Roundup-Ready soybeans went from about 15 million tons in 1996, before their introduction by Monsanto, to an expected 50 million tons in 2023, a dramatic increase.[3] Where there once grazed beef-producing Angus cattle, there now grow GM soybeans. The cattle, meanwhile, have been moved into feedlots. It is estimated that about 70 percent of all Argentine cropland is now dedicated to soybeans.

The economy of Argentina is increasingly dependent on soybean prices paid by China, which feeds the imported beans to pigs for domestic pork consumption. This soybean-focused model nudges Argentina back into the classic "extractive" economies of developing nations, such as petroleum does for Nigeria or diamonds for Congo. Extractive economies depend on income from one predominant source of exports, a risk that has been called "the resource curse." When world prices rise for oil, diamonds, or soybeans, the economies of the countries grow. When the prices drop, their economies destabilize.[4]

A second problem with herbicide-resistant soybeans monopolizing the acreage of planted crops is the increased use of Roundup or its generic version, glyphosate. Even before the demand for it grew worldwide due to the introduction of Roundup-Ready crops, evidence was being published that the herbicide had ill-health effects on the humans applying it. As acreage of the GM crops grew, so did use of the herbicide. In her 2017 book *Whitewash: The Story of a Weed Killer, Cancer, and the Corruption of Science*,[5] Carey Gillam describes reports of birth defects and cancers in communities living close to croplands sprayed with glyphosate. Monsanto denied any linkage between the herbicide and such health issues. A deeper dive into the literature only reveals confusion and contradiction.

Third, herbicide-resistant GM crops are so profitable that planters grow them at the risk of losing farmland biodiversity, which leads to vulnerability to new pests. Think Perú and Ireland. If a sole crop takes over all fertile land, infection by a deadly pathogen could wipe out the entire harvest and lead to an ecological and financial disaster. Unless carefully regulated, patented monocrops may harm society by increasing our dependence on a single source of foodstuff.

The farmers of the Global South had developed, ever since pre-Columbian days, a large variety of seeds that were finely adapted to microclimates and soils. Contemporary South American farmers have now replaced these ancestral seeds with high-yield patented Roundup-Ready seeds. Were a nasty pest to come by and wipe these out, we'd be back in Ireland, circa 1850. To mitigate against such a catastrophe, Norway has established a Global Seed Vault on the island of Spitsbergen in the remote Arctic Svalbard archipelago. The Seed Vault contains more than a million distinct crop samples, representing 13,000 years of agricultural history. It is there in case anyone on our planet ever needs to replenish a crop that has been wiped out.

The risk of biodiversity loss, the economic dependence on one predominant export source, and the uncertain health effects of too much glyphosate weed killer—all on the altar of economic development—are unintended consequences of the monopolization of harvests by patented GM crops. Setting aside the IP maneuvers we discussed in chapter 18, we can at least understand the delicate balance that exists in the pharma industry between patented drugs and their temporarily high prices. It is part of a societal deal intended by all players: the benefits to pharmaceutical innovation are worth the patent system's costs. Yet the unintended consequences that come out of herbicide-resistant GM crops feel more like a Faustian bargain than an informed societal agreement.

But wait. There's more.

## Litigating Farmers

From the moment that sales of patented herbicide-resistant crops started going up, Monsanto established so-called Technology Agreements that farmers had to accept as part of buying Roundup-Ready seeds. These contracts said that as an agreed-upon condition of using the seeds, the farmer had to come back to Monsanto's dealers every year for more seed. Many US farmers could not believe that these Technology Agreements would hold up in court. They thought Monsanto was overreaching and forcing them to sign up for things that their ancestors on the family farm would have never agreed to. You could almost hear the dialogue between Monsanto and the farmers.

"We should be able to save seed from one planting season to the next and replant it with no problem," said the farmers. "We don't have to go back to Monsanto, no matter what Monsanto says in these contracts."

"Well," answered Monsanto, "The Roundup-Ready seeds are patented and that allows me legally to restrict you from saving and replanting them."

"Patents on seeds? What cockamamie idea is that? Whoever heard of such a thing? And even if these seeds are patented, doesn't the first sale of a patented good 'exhaust' the patent? If I buy a patented pen, I can do whatever I want with it, no? You can't charge me a yearly user's fee. And so it is with these seeds. I can save them and replant them year after year. I don't have to pay you again and again."

The farmers may not have known what Monsanto could do with its patents, but Monsanto was more than happy to enlighten them, and the hard way. It is estimated that Monsanto filed close to 120 lawsuits against farmers in the decade between 2003 and 2013.

"For starters, pens are not seeds," replied Monsanto. "Pens do not self-replicate like seeds. Even the first sale of a patented pen doesn't allow you to make additional copies and sell them in competition with the owner of the patent. The first sale doctrine does not go that far. Your payment for a bag of seed buys the seed for one planting season. Seed harvested from the first season and saved for replanting is new seed. It is still patented and the sale of the bag to you the previous year did not exhaust the patent rights."

The protestations of the farming community were heard all the way to the Supreme Court. And Justice Elena Kagan, writing for a unanimous court in *Bowman v. Monsanto,* a decision handed down in 2013, sided with Monsanto and against the farmers.[6] The syllabus of the opinion put it succinctly: "By planting and harvesting Monsanto's patented seeds, [farmer] Bowman made additional copies of Monsanto's patented invention, and his conduct thus falls outside the protections of patent exhaustion. Were this otherwise Monsanto's patent would provide scant benefit. After Monsanto sold its first seed, other seed companies could produce the patented seed to compete with Monsanto, and farmers would need to buy seed only once."[7]

This is now the law of our land. The first sale of patented replicating live beings—whether they are Roundup-Ready seeds, *Phaffia* yeast cells that color salmon red, or transgenic cows that produce human hormones in their milk—does not exhaust the patent rights. You buy and pay for the first generation of these patented beings. You can then use them for their intended use, like extracting hormones from milk or feeding yeast to salmon. But beware of making additional copies of the patented yeasts or enticing the patented cows to make more of themselves at twilight out there in the Pampas. If you turn down that road, you may go into competition with the patent owners and run into serious trouble.

I am not saying that the Supreme Court was wrong; the decision makes sense. But it sure ticked off the farmers. Because of the Roundup-Ready litigations, Monsanto developed a fierce reputation in the US farming community, along the lines of "Don't mess with Monsanto unless you're ready to go to court."

Nowhere was this more evident than in a lawsuit brought in 2013 by a large group of organic farmers against the company.[8] The plaintiffs were concerned that their fields might become inadvertently contaminated by trace amounts of patented Roundup-Ready seeds blowing in from neighboring land. Such an event would undermine "organic" certifications, which ensured that plants from their farms did not include more than trace amounts of GMOs. Plus, fearing the worst from an aggressive Monsanto, the organic farmers did not want to be sued for patent infringement either. They sued first. A more accommodating Monsanto, however, had already assured the world in its website that it would never sue if the amounts of Roundup-Ready seed were less than 1 percent of a total bag. The Court of Appeals held that Monsanto's assurances were enough and sent everyone home. All along, the answer was blowin' in the wind (and could be found on Monsanto's website).

Both Roundup and Roundup-Ready soybeans are now generic and in the public domain. A US farmer can save and replant all the generic seeds she wants, without having to go back to anyone for more.[9] In 2016, a year after the patent on Roundup-Ready soybeans expired, Monsanto, in a classic evergreening maneuver, announced new and improved Roundup-Ready soybeans, which it calls Roundup-Ready 2 Xtend® and which, of course, are patented. The seeds contain what is called a pair of "stacked traits," that is, resistance to the original Roundup on top of resistance to a different herbicide called dicamba.

In 2018 the German multinational company Bayer AG acquired Monsanto for more than $60 billion. And, while the Monsanto name has disappeared, the trademarks of its famous products have not. Just as Pfizer's trademark Celebrex lives on after its patents on celecoxib have expired, Monsanto's Roundup-Ready name, which is common to the earlier generation of now-unpatented soybeans and to the new generation of patented ones, lives on.

As is often the case with patented pharma drugs, there is now competition from the generic seeds, although they are only available with the one original trait, single resistance to Roundup. Unless the double resistance soybeans offered by Bayer are worth a premium, the prices for the new Roundup-Ready 2 Xtend beans should not be as high as the single resistance seeds once were. But, even if prices drop, the unintended consequences of mono-cropping herbicide-resistant GMOs will still be there. Until more government oversight regulates the non-price aspects of herbicide-resistant GM crops—including risks to health, biodiversity, and to well-balanced economies—it is not clear to me that the public has fully bought into this agribiotech patent bargain.

My take on Monsanto is that it made a strategic marketing blunder from the get-go. Its first GM crop, Roundup-Ready soybeans, favored increased sales of Monsanto's own herbicide. It also favored farmers, in that the selective killing of weeds was a profitable thing. What it did not do was directly favor consumers. Why should you as a consumer care that Monsanto could sell more Roundup, or that soybean farmers could make more money? What was in it for you? Had Monsanto come out with some good GMOs, such as a legume enriched in vitamins, or a transgenic wheat with a lower glycemic index that benefits diabetics, consumers may have felt that GMOs aren't so bad after all. Such a strategy might have given consumers the feeling that the company had their interests at heart and wasn't pushing genetically engineered food only to revive a weed killer that was losing market share. Yet hindsight about a public relations misstep would not cure the unintended consequences that came out of Monsanto's biggest impact: the steady monopolization of the world's croplands by herbicide-resistant GM crops. Monsanto is no more, but its confounding legacy lives on.

My law professors of forty years ago did not yet have a full picture of what "monopolization" means in the modern world. They thought of controlled and acceptable monopolies, leading to the temporary high prices of patented drugs. They did not know that patented soybeans resistant to weed killer were fifteen years in the future and would cause havoc with the world's biodiversity. I wish I could go back to law school today and argue about Faustian bargains; at least until I was told to sit down.

Before we leave part IV, let me disabuse you of any notion that I am not in favor of patenting GM crops. I am in favor, especially if the patents protect crops that benefit consumers, not only patent owners and farmers. Take the ability to confer drought tolerance to food crops. Severe weather-related events such as lengthy droughts are no longer uncommon in this era of climate change. Inventions of drought-tolerant crops are therefore critical to assure food security. Investing in such work, which includes receiving approval for commercialization, are urgent—and expensive—matters. And, as we have seen, without some period of exclusivity, private investment dwindles.

I have gladly helped several Argentine entities obtain patents on ingenious drought-tolerant plants.[10] To make a plant like wheat tolerant to drought, scientists have inserted into it a gene from sunflowers. Sunflowers not only look like the sun, their yellow petals gushing outward from the center as so many rays, but they love sunlight and cannot get enough of it. Young sunflower plants have evolved a marvelous biological mechanism that keeps them pointed at the sun, turning, and following its daily path across the sky. If you are a flower, keeping warm all day long seems to help attract pollinating bees.

It has long been known that sunflowers are better tolerant to drought than most other crops. Therefore, supported by a mix of public and private investments, Dr. Raquel Lia Chan, the director of the Agrobiotechnology Institute of Santa Fe, Argentina, and her team identified and extracted several genes responsible for the sunflowers' tolerance. They genetically engineered one of those genes into wheat, and the reprogrammed wheat became tolerant to drought and had higher crop yields than the unprogrammed one.[11]

A client of our firm, Bioceres, a startup Argentine agribiotech company, took international licenses to Raquel Chan's patents and is commercializing drought-tolerant crops around the world. I am pretty sure that without patent protection, the investments would not have occurred, and humanity would not have such promising, life-sustaining products at its disposal. I hope that having learned lessons from the Irish Lumper and the Roundup-Ready eras, governments will insist that drought-tolerant crops be used in ways that will not only assure plentiful food but also preserve biodiversity.

# Part V

# The Path Running Alongside

The dissenting voices I heard over the years led me to a heightened understanding of the careful balancing act needed to create and maintain an evenhanded patent system. Late in my career, they inspired me to redirect part of my legal practice onto a path that runs alongside that of my earlier times. The path reflects efforts to use the patent system to bring economic and social improvements to the lives of disenfranchised Indigenous communities.

# 22

# *Tikkun Olam*

After the ignominious collapse in 2001 of the lawsuit brought by forty pharmaceutical companies against the government of South Africa that we discussed in chapter 20, I became increasingly conscious of some of the sketchy aspects of the international patent system. These thoughts evoked a Hebrew concept known as *tikkun olam*, which means "to repair the world." It is the idea that we all bear responsibility not only for our own moral, spiritual, and material welfare but also for the welfare of society at large.[1] The more I mused about it, the more I felt the need to repair the world—at least my world, the one of intellectual property.

## Reverse Tech Transfer

I am not a legislator with an established path to try and bring about global solutions. I am an attorney, and if I wanted to repair the world, I would have to do it one case at a time, as is the practice of attorneys. I would have to choose the right cases and establish some precedents, or at least some examples. Motivated by the overlap of human rights and intellectual property that had changed the conversation in the South African litigation, I decided to establish a pro bono practice in my law firm to focus on the type of case that would allow me to repair the world one invention at a time. Pro bono work has a long and honorable tradition in the legal profession. It is work done free of charge to help underrepresented clients who are unable to afford a lawyer. While it has been applied in as diverse areas as assisting criminal defendants or aid the elderly, I decided to mix pro bono work with IP law.

Around 2010, at American University Law School, I learned from Professor Olivier De Schutter, who at the time was the United Nations special rapporteur on the right to food, of a little-known realm of human rights: the so-called Economic, Social and Cultural Rights (ESCRs). In 1944 Franklin Delano Roosevelt, in his State of the Union address, described ESCRs as a "Second Bill of Rights"—the first bill being political and civil rights, such as freedom of speech or religion.[2] The ESCRs, in contrast, deal with economic rights such as jobs, development, health, shelter, and food. While the right to own IP is not a recognized human right, the right to benefit from one's creations is an ESCR. The 1976 United Nations' International Covenant on ESCRs recognizes "the right of everyone . . . to benefit from the protection of the moral and material interests resulting from any scientific, literary, or artistic production of which he is the author."[3]

I envisioned that our firm's pro bono practice would be inspired by the idea of redeeming ESCRs through intellectual property. The practice would help disenfranchised groups of the world gain commercially important IP and use it to advance their rights. We would represent Indigenous communities and their members, who may have created a "scientific . . . or artistic production" that could be protected and leveraged, so that their villages or tribes could benefit. The idea was simple: allow impoverished and underrepresented groups from the developing world to use the rights and benefits of the patent systems of the developed world to benefit from economically industrialized markets. This would be a sort of "reverse technology transfer." Instead of technology being transferred from the Global North to the Global South with funds heading North, as has been the traditional case, the flow would be reversed: technology from the South would establish itself in the North, with funds flowing South.

The reverse tech transfer approach would not depend on there being patent systems in the countries of origin that are like the ones in the North. That is a common complaint from the patent holders in Europe and the United States and was one of the main reasons for the 1995 TRIPS agreement we discussed in chapter 20. All a country or community would need to learn (from our pro bono practice or others) is how to use the systems of the North.

Two of my pro bono cases are worth a tale. In this chapter I will tell you of the story of a blue fruit juice from a Colombian rainforest tree. In the next I will tell you about feeding leafy vegetables though their leaves.

## Tattoos

In the Chocó rainforest of Colombia, on the Pacific side of the Andes (reaching all the way to the Peninsula Darién in Panamá) there live Indigenous peoples who go by the names of Emberá, Waunan, and Kuna, who, to keep it short, I will call Emberá.[4] They inhabit one of the richest biodiversity hotspots on Earth. These Indigenous groups were originally seminomadic hunter-gatherers who had lived in the area since at least the sixteenth century. Afro-Colombian communities have also been in this territory since colonial times, when they were brought as slaves for gold and platinum mining. More recently, modernism has intruded. The construction of the Pan-American Highway, the appearance of mechanized illegal mining, large-scale deforestation to build new farming establishments, and the appearance of narco-guerrillas have shaken up the historical equilibrium. Many of the original communities have lost their precious forests and have become subsistence farmers or informal employees of these new activities.

One of the Emberá joys and talents is body painting. For as long as anyone has documented, they have used the dark blue juice of the fruit from the *jagua* tree (named in honor of the jaguar as a sacred animal) to decorate themselves for rituals, ceremonies, or just for fun. They may use the juice directly or after pretreatments. Their tattoos and colorings remain stable for quite some time. Figure 22.1 shows a tattooed Emberá girl.

Earlier this century, a privately funded Colombian company, Ecoflora Cares, working with an organic chemist from a local university in Medellín, extracted the active ingredient from the blue juice of the jagua

Figure 22.1. An Emberá girl tattooed with jagua juice. Source: Wikimedia Commons. Author Yves Picq, http://veton.picq.fr. Licensed under the Creative Commons Attribution-Share Alike 3.0 Unported, 2.5; 2.0 and 1.0 Generic licenses.

fruit and, through a series of chemical reactions, made it into a stable, free-flowing powder. The powder has a beautiful cobalt-blue color, which shows as dark in figure 22.2.

Ecoflora Cares wished to commercialize the powdery blue color. There is a lot of demand for edible blue colors, and the company figured it could capture some of the worldwide market. Think blue edibles like M&M's, cake frostings, or Viagra. The food industry has referred to blue as the "missing holy grail." Blue is scarce in nature, and hence there is a dearth of acid-stable and safe blue coloring for foods. This is especially the case for adding blue coloring to carbonated drinks, such as sparkling water. The pH of such drinks is in the range of 3 to 4, acidic enough to cause breakdown of most of the existing blue additives. So, the food industry has been seeking a stable blue that would have a long shelf life in soda water and other beverages. Ecoflora's jagua blue powder meets the

Figure 22.2. A cut-open jagua fruit and Ecoflora's blue powder. Courtesy of Ecoflora Cares.

requirement in that it does not easily degrade at the pH of carbonated drinks.

The company's management was socially conscious and wanted to market their blue in a manner that would respect the sustainability of the fruit and benefit the Emberá sourcing communities. They wanted to follow the strictures of the Convention on Biological Diversity and its accompanying Nagoya Protocol (CBD/Nagoya). These are international treaties for the protection of rainforest genetic resources and for access and benefit sharing with the communities that cultivate them.

The Convention on Biological Diversity, also known as the Rio Treaty, was put in place in 1992 as an attempt to stop the practice of "biopiracy." Until the CBD was signed, it was the unchallenged custom for prospectors from the Global North, such as academic or commercial researchers, to freely explore the ecosystems of the South. They would find, extract, and take home with them valuable natural resources, such as medicinal plants. They would then commercialize them or, after further modifications, patent and commercialize their derivatives. The prospectors assumed that the rainforests of the world belonged to everyone and that no permission was needed to access them or to take their valuable resources.

The story of the neem tree is a striking example in that the discoveries happened two years *after* the United States signed the Rio Treaty. This sorry tale involves W.R. Grace and Co., of Columbia, Maryland. You may recall the guano fertilizer exports from Perú that we discussed in chapter 21. The guano carried a nasty pest that was the cause of the Irish potato famine in the mid-nineteenth century. Well, none other than the Irishman William Russell Grace, known to the world today as "W.R. Grace," escaped the very famine to, of all places, Perú. In Lima, in 1854, together with his brother Michael, W. R. founded Grace Bros. & Co. Cluelessly, their company kept exporting contaminated guano to Europe. Yet these guano exports are not the only ill-informed misstep the company took during its early life. In 1994, 150 years after the guano gaffe, W.R. Grace received a European patent on methods of controlling fungal infections in plants with a composition that included extracts from the Indian neem tree. In 2000 the patent was opposed and revoked on the basis that the fungicidal activity of the neem tree had long been known in Indian traditional medicine.[5] Talk about serial blunders.

The Rio Treaty was enacted in 1992 to try and stop, or at least regulate, practices such as obtaining and patenting products that arise out of traditional knowledge. The treaty says that sovereign nations own and can control access to their "genetic materials" and that they have "the sovereign right to exploit their own resources."[6] "Genetic material" is defined broadly to mean "any material of plant, animal, microbial or other origin containing functional units of heredity." In simple words, if a material has genes, the country of origin has control over it. The Nagoya Protocol of 2010 followed the adoption of CBD. It reinforces the obligation to negotiate for access to genetic resources and to establish arrangements for sharing financial and other benefits with the local populations. The basic principles of CBD/Nagoya are conservation, sustainability, proper access to, and benefit sharing with, any relevant Indigenous communities.

While the United States signed the treaty, our Congress has never ratified it. It is not the law of the land. In fact, the United States hasn't even *signed* the Nagoya Protocol. It might not surprise you that the US pharmaceutical lobby, supported by a libertarian conservative block in the Senate, strenuously opposed ratifying the Rio Treaty in 1992.[7] Their objections were partly based on an incorrectly perceived notion that the industry would have to share with countries of origin any IP obtained from their genetic resources. Add to this the aversion that, as a matter of principle, conservative politicians had toward international treaties, fearing that they would make the United States lose its sovereign rights. The sum of these opponents blocked ratification.

I have since had a recurring fantasy: One of those conservative politicians is visiting Yellowstone as a tourist. He sees a group of Brazilian, white-coated scientists busily collecting water from one of the marvelous, orange-colored hot springs near Old Faithful. When he asks, the Brazilians explain that these springs are rich in so called extremophilic bacteria, which thrive in these hot acidic homes. "Imagine the interesting things we will learn from studying these microbes," they add. Would the politician's conservative instincts encourage the Brazilians to keep bioprospecting and wish them well? Or would he say that the US has sovereign rights over the bacteria and that they need permission before taking their test tubes back to Rio? I am pretty sure what his answer would be. Political hypocrisy would then extend all the way to the hot springs of Yellowstone.

In 2011 Nicolas Cock Duque, the CEO of Ecoflora Cares, learned of our law firm's pro bono program through Public Interest Intellectual Property Advisors (PIIPA), a Washington DC–based nongovernmental organization (NGO) that connects IP pro bono lawyers with potential clients worldwide.[8] Cock Duque came seeking help to obtain patents on the blue powder and its applications. When we investigated the Ecoflora Cares blue color proposal, we knew that we had found a worthy project on which to test our ideas.

While the development of a stable blue powder was not the invention of any one member of the Emberá, the juice of the jagua fruit and its use in body painting did originate with the tribe. Protecting Ecoflora's invention of a stable blue powder (a patentable improvement over the natural jagua juice) also allowed us to leapfrog over another issue that often arises with direct representation of a tribe. Tribal property, including collective know-how (which is IP by another name), is treated communally, not individually. The patent systems of the world, however, require the naming of individual inventors, which is a problem when one deals with communal IP. Thus, focusing on a downstream invention, made by a university chemist, and owned by a private company, simplified matters.

I liked that Ecoflora Cares was conscious of Rio/Nagoya. It worked with a group of Colombian national and local governments, creating a business and regulatory network that would allow them to ethically source the fruit. They could then develop the blue powder as an additive for foods, drinks, and cosmetics. Ecoflora entered deals with several local communities to establish the production and supply of the jagua fruit with commercial partners. Through a benefit-sharing agreement, community-owned fruit suppliers would share in monetary and nonmonetary benefits of any commercialization of the blue color. Additional benefit sharing would occur through nonprofit organizations that train local producers on sustainable sourcing. Ecoflora Cares was also a member of the Union for Ethical BioTrade, an Amsterdam-based organization that certifies that if rainforest genetic resources are included in commercial products, they are used in accordance with the principles of Rio/Nagoya.

Our pro bono team felt that representing Ecoflora as a proxy for the tribe was a worthwhile endeavor. We rolled up our sleeves and went to work, all of us intent on repairing the world.

## Beads and Mirrors

Our law firm's pro bono practice has since obtained for Ecoflora Cares two major patent families on the jagua color technology. The patents cover the blue-colored powder, its detailed chemical composition, its manufacture, and its use in the production of consumer products, such as solid and liquid foodstuffs, personal care goods, or medicines. Our client now has patent protection in many diverse regions of the world. It has patents in tropical places in the Global South where the jagua tree might grow and where the patented production methods might be used (such as Brazil, Perú, and Costa Rica); and places in the Global North where the blue color will go into foods and drinks (such as the United States or Europe).

The fact that the United States has neither ratified the Rio Treaty nor signed the Nagoya Protocol did not mean that I would ignore what seem like very worthwhile international agreements. To let the world at large know that Ecoflora will commercialize a blue powder derived from a rainforest genetic resource, I voluntarily included in its issued US patents a "Statement of Access and Benefit Sharing." The statement explains that the sourcing and commercialization involving this resource comply strictly with the principles of the Convention on Biological Diversity. The Patent Office had never seen anything like this. But I insisted, and the examiners in charge added it.

In parallel to our IP efforts, Ecoflora started extensive testing for submission of safety information to the FDA and other international regulatory authorities. In 2023 the FDA deemed the blue powder safe for human consumption and in 2024 issued a license to include it in edibles, cosmetics, and medicines.

Ironically, filing and obtaining patents for Ecoflora, and testing for safety to get approval by the health agencies, turned out to be the easier parts of the project. It was harder to convince some of the major food additive companies of the world to partner with Ecoflora to bring blue foods and drinks to the markets. It is not that these companies were skeptical of adding blue to foods; as I have explained, edible blues are the holy grail of the food industry. Neither did the companies have any problem with approval by regulatory agencies, with the patent protection we had obtained for our client, nor with compliance with Rio/Nagoya. Their problem was, let's say, cultural.

The "reverse technology transfer" idea was a novel concept that many had not previously encountered. The idea of paying for Colombian Indigenous-derived technology was foreign to them in more ways than one. Let me illustrate this with an anecdote: In 2014, together with an officer of Ecoflora, I traveled to Europe to negotiate a deal with a multinational company having an interest in bringing this blue to the food industry. After introductions and a customary exchange of pleasantries, we got down to business. I was immediately worried when I saw that the company's team did not have a patent attorney. In fact, there was no one from the legal department there. And it was clear from the get-go that we couldn't have been any more far apart in our expectations.

The multinational's approach, which I will generously call "neocolonial," was an offer to buy all the jagua fruit Ecoflora could sell it; it would then take care of the rest. Maybe it would pay one single fee for the patents but nothing much more than that.

When I heard this, I let them know how I felt through my barely disguised expression of shock. "Fruit?! You want to buy fruit?" I said. "We didn't come all the way from Washington and Medellín to sell fruit."

## Hemming and Hawing

"We came here to sell intellectual property," I added, calmly. "You know . . . patent rights? A worldwide license with upfront payments, milestones, running royalties, that sort of stuff? I'm sure you know what IP is. Last time I checked you had several thousand patents around the world. And many IP deals."

Silence.

"Of course, we'd be happy to sell you all of your yearly fruit requirements," interjected my client. "However, if you don't want to deal in IP, we'll sell you the fruit but offer the patent license to one of your competitors. Then you'll need to negotiate with them to add the patented blue powder to foodstuffs."

Sorry, said the company representative, but he had no authority to negotiate such a deal. And it didn't seem too profitable to do it our way. That pretty much ended the negotiations.

The whole thing reminded me of the way European travelers would go to a tribal shaman and, in exchange for the leaves of a tree that he used on sick people, give him some beads and mirrors. With international legal IP advice, such as from a lawyer of our pro bono group, things have changed. IP is a great equalizer, and, when added to the norms of Rio/Nagoya, has made it more equitable for underrepresented Indigenous communities such as the Emberá and their commercial partners such as Ecoflora to gain negotiating power they never had before.

We encountered the same skepticism many times. In 2017 our client decided that it would try to get the multinational food companies to compete. So, we organized a worldwide virtual auction, using our firm's computer servers in Washington, DC, as an extranet, where we placed several dossiers of information. After the companies paid an access fee and signed nondisclosure agreements, we gave each a unique password, and they received access to several databases. One was a folder full of regulatory information, another a folder full of patent information. Other folders had trade secrets on the process of production, a template supply and licensing agreement, and spreadsheets with projected business models.

Ecoflora received twelve companies' offers of interest. Several of them discussed possible agreements, and one, Chr. Hansen of Hørsholm, Denmark, who was not the neocolonial one that wanted to buy fruit, eventually signed a deal. The Chr. Hansen division that did the deal now goes by the name of Oterra. The contract is a supply and IP licensing agreement under which Ecoflora Cares will receive compensation for the sale and distribution in the United States of the blue powder for coloring foods and beverages. The legal and contractual groundwork laid under Rio/Nagoya assures that the Emberá will receive benefits. There have even been preliminary ideas to build a plant onsite that would allow community members to collect the fruit, extract the juice, and add value locally within the process to yield the stable blue powder. Such a project, by providing steady employment and income, could move the Emberá up the distribution chain, establishing additional incentives to protect and grow the jagua tree while preserving and defending their territories.

The Oterra deal is the culmination of almost ten years of legal and negotiating work by a team made up of our firm's pro bono group and our client. It is premature to conclude if the big project has been a success. An undertaking like this, which aims to assist Indigenous communities in navigating the treacherous currents of the modern commercial world, is built one brick at a time. Each brick is a significant, although smaller, success: establishing a proper framework under Rio/Nagoya; obtaining allowance of patent claims; obtaining approval from health authorities to add blue color to carbonated drinks and other foodstuffs; finding the right multinational company to do the deal (in particular, a company willing to play by the rules of fair trade and benefit sharing); and negotiating and executing a deal that respects IP while also doing good for the rainforest communities.

It is too early to tell if the brick foundation that we have helped our client build will be enough for the big success: improving the lives of the Emberá. Will the project create steady gains in employment and in the ESCRs that flow from it—health, education, food, shelter, and economic development? Will it help preserve the rainforests and restore tropical ecosystems through replanting of more jagua trees? Have we helped repair the world? Time will tell.

The Emberá don't really care whether their drinks are colorless or blue, sparkling, or plain. If I had to venture a guess, I would think they prefer their liquids clear and flat. They probably shake their heads at the notion that we in the United States or Europe would spend money to buy a blue fizzy cold drink to "hydrate" properly after a yoga session. Maybe someday, when they see the benefits that flow to them from the sale of blue soda drinks in New York or Paris, they will—fortunately or unfortunately—care. Such are the ways of the modern world.

Now on to leafy veggies. While the Indigenous community involved, the Monacan Tribe of Central Virginia, is not in the hemispheric South, it is easily placed in the Global South: underrepresented and impoverished.

# 23

# Foliar Feeding

The inventor in this story is Lucas Swampdog Tyree. His Native American name is Shungnashkinyan Wiwila Kapapa, literally "Swamp Dog." I first met Lucas, a member of the Monacans, in Lexington, Virginia, as he drove up at high noon in his Harley: helmet and ponytail to the wind. It was a very hot day, close to the Fourth of July 2015, and Lexington was celebrating with a Main Street parade. Not a cloud in the sky, not many trees to hide from the oppressive sun, no breeze. There were lots of children in patriotic regalia, parents pushing strollers, men in Bermuda shorts, many American flags, and even a few Confederate ones. Lucas had told us, my wife, Sandy, and me, about his motorcycle, so we recognized him instantly as he slowly pulled up to the agreed-upon corner. He didn't make a noisy entrance into Lexington, wise to the historical fact that his family, the Monacan Tyrees, had once been classified as "mixed blooded negroes." That way, Virginia could better discriminate against them under Jim Crow.

Over the course of our acquaintanceship of several years, Tyree would periodically tell me details of the hideous inequities that he and his family had endured. I learned of Dr. Walter Plecker, a physician who, from 1912 to 1946, was the registrar of vital statistics of the state of Virginia. Plecker made it his life's work to assure that Native Americans, like the Monacans living in the state, would not try and pass off as "Indian." If they did, they might take improper advantage of the rights and privileges not due them if they were instead classified as "Negroes."

Plecker wrote a notorious letter dated January 1943 addressed to "Local Registrars, Physicians, Health Officers, Nurses, School Superintendents, and Clerks of the Courts."[1] In the letter, which he sent to his "Dear Coworkers," he warned of Native American parents who he believed might be falsifying their children's racial designations so as not to be blocked from receiving public benefits. There should be no schooling, health care, or intermarriage for them; Virginia considered them to be Negroes. In an attachment, listed by county, he included the surnames of families he designated "mongrels."[2] The Tyrees of Amherst, Alleghany, and Campbell Counties are listed among those who were trying to "escape from the negro race" by stating "Indian" as the race of the child's parents.

In a particularly venomous excerpt, oozing the contempt that the racist doctor had for Native Americans, he talks about "these people," who were sneaking their way out of being Black: "Now that these people are playing up the advantages gained by being permitted to give 'Indian' as the race of the child's parents on birth certificates, we see the great mistake made in not stopping earlier the organized propagation of this racial falsehood."

Plecker ends his letter with a flourish, warning that many other "mulattoes in Virginia are watching eagerly the attempt of their pseudo-Indian brethren, ready to follow in a rush when the first have made a break in the dike." Plecker's attempt at literary swagger comes across as more pathetic than poetic.

In helping me prepare this chapter, Tyree also told me of brutal aggression by the Ku Klux Klan, which ran his family off its ancestral lands by violence. He wrote, "My family survived, though the acreage decreased slowly over time as the US Forest Service condemned our lands, racial violence against us forced my ancestors to flee and sell their land for pennies, and racist policies kept them from employment or education." The contemptible bigotry of Walter Plecker and his kind, he added, "effectively stole my family's potential for decades."

## Redemption through Creativity

Lucas was intent in righting all these wrongs. He was the first of his family to graduate from high school. He didn't stop there: He went on to college at the University of Hawai'i and then received an MS in environmental science from Yale. His professors wanted him to stay for his PhD, but Lucas was drawn back to his ancestral lands. He went to live with his father in the Blue Ridge Mountains near Vesuvius, Virginia, to work the land and improve his tribe. He had been referred to me by the Native American IP Enterprise Council (NAIPEC), an organization whose aim was to help Native Americans improve their lives. NAIPEC introduced him as the inventor of some interesting technology, and would we represent him pro bono so he could get some patents? His personal story screamed "human rights violation" so loudly that it didn't take much thinking to bring him on as a client.

On that hot July day in 2015, my wife, Sandy, and I drove down to Lexington, Lucas drove up from his farm near Vesuvius, and we met for lunch. Right around the corner from the restaurant, behind Main Street, is the campus of Washington and Lee University—where Robert E. Lee, his whole family, and even his horse Traveller are buried. Before catching up with Lucas, Sandy and I visited the peaceful campus and Lee's Chapel. Watching all that Confederate paraphernalia, I had no doubt that poisonous Plecker thought more highly of Lee's horse than of Monacan children. Surely, Lucas knew of Lee's ghost floating around the all-white parade. But he said nothing about Lee or the Confederate flags.

He was a classic inventor: smart, curious, optimistic, and focused on the future. Over hamburgers, he told me that he had invented a method of feeding plants by spraying their leaves, especially big-leafed plants such as lettuce. He had developed a formula that contained all the nutrients that would normally be absorbed from the soil. To facilitate foliar absorption by spray bottle, he had added a few extra components that made it possible to give such plants all they needed, without involving the roots. The plants could then grow in the clean mountain creeks of the Blue Ridge, in the eroded soils of Haitian mountainsides, or in the sands of Nevada Indian reservations. He dreamed of establishing foliar feeding schools, where he would train young students from Indian Country to become self-sufficient farmers. He thought that his invention would also work in so-called vertical agriculture centers, which are basically hydroponic farms set up in recycled urban buildings.

Figure 23.1 shows Lucas Tyree in his family's land near Vesuvius, Virginia, in front of the mountain that his fifth great-grandfather fought for in the War of 1812 and that Lucas repurchased after it was lost/taken.

A few months after the Lexington lunch, our pro bono team filed a patent application for Lucas. While his application was pending, I discussed Lucas's invention with the chief of research of one of the largest fertilizer companies in the world. He said that, if the method worked, it would be the "Holy Grail" of plant feeding. It was technology that had been sought after for many years. Between the holy grail of blue-colored foodstuffs, and the holy grail of foliar feeding, my pro bono practice, originally aimed at redeeming ESC rights, seemed to have grown sacrilegious.

For several years Lucas's invention was rejected by the Patent Office as "obvious" in view of prior publications that portray foliar feeding as supplementing the regular growth of garden plants by spraying their leaves. The Office also insisted that the scope of any patent be limited to lettuce. We replied that Lucas's formulation is not a supplement but serves as the plant's primary nutrition, especially when the roots are not in contact with any nutrients at all. We also argued that lettuce is one example but that the invention extends to all leafy vegetables. In 2024, as this book was going to press, the Patent Office relented on both issues. As a result, Lucas will receive broad patents on foliar feeding all leafy veggies.

As I was writing this chapter, Lucas spoke some more of his quests. He was proud of his family and of his own achievements: buying back the lost acreage where he now lives, educating himself, and inventing his patentable formula. What struck me most was his sense of reckoning. He felt that he had eventually defeated Plecker and his ilk. His triumphs were his best revenge. But it was more than that: his comments transcended personal accomplishments. They were a reflection on how all of us in society suffer when we discriminate and suppress those whom we consider inferior or different. That brought me back full circle to my early years.

Tyree's feelings evoked that day in 1984 when I was witnessing the inauguration of the Wellman Building at the Mass General Hospital

Figure 23.1. Lucas Tyree in his ancestral lands in Virginia. Courtesy of Lucas Tyree.

Department of Molecular Biology, which I described in chapter 3. I again heard the chairman of Hoechst AG, who had flown over from Germany to cut the ribbon, reflect on a similar reckoning. He commented on how German science had declined as it vanquished its Jewish scientists during Hitler's regime. And how, since Germany had brought scientific ruin on itself, Hoechst had to come to the United States to learn modern biotechnology.

Although the Germans atoned and Plecker never did, I recognized a comparable sense of justice playing out in both stories. In the end, the tormentors lost, and inventiveness was the reward given to their earlier victims.

No one can keep down the creative spirit forever. The evildoers of the world may try to suppress her, but the muse invariably returns and breaks free. As Stephen Sondheim poetically says in his musical *Follies*, "Good times and bum times, I've seen them all / And, my dear, I'm still here."

## Small Is Beautiful

The Emberá blue color project and Lucas Tyree's foliar feeding methods are what are known as "intermediate technologies." British economist E. F. Schumacher coined the term in his 1973 book *Small Is Beautiful: Economics as if People Mattered.*[3] The term refers to technologies that, while not as high-tech as genetic or electronic engineering, would create more environmentally friendly employment in developing nations. The Encyclopedia Britannica Online defines "intermediate technologies" as consisting of "simple and practical tools, basic machines, and engineering systems that economically disadvantaged farmers and other rural people can purchase or construct from resources that are available locally to improve their well-being."[4]

Because of our law firm's concept of helping underrepresented and impoverished Indigenous communities with their intermediate technologies, in 2015 the *Financial Times* of London gave us the Most Innovative North American Lawyers Award for Innovation in a Social Responsibility Project" We were in head-to-head competition against the pro bono practices and projects of several large North American law firms. Much to our joy, we beat them.

In a short thank-you speech I gave when accepting the *Financial Times* award, I talked about the responsibility we bear to repair the world. I reminded the audience of IP lawyers that they could repair the world even if it was one patent at a time.

# Epilogue

Half a century has elapsed since Paul Berg did his pioneering DNA recombination experiments. During that span, the worlds of biology and IP law merged, and both worlds changed forever. And, while not perfect, the transformation has been good for society.

The patent system is tailor made for the young startups that sprung up in the wake of the inventions of genetic engineering and monoclonal antibodies. Creative companies like Sanaria, Ecoflora Cares, or Bioceres, and individuals like Lucas Tyree, need well-guarded beachheads to give them time to grow and bring their vaccines, natural blue colors, drought-resistant crops, or foliar feedings to the world. All they have at the beginning of their life is intellectual property. Giving them a temporary foothold and watching them succeed was one of the great pleasures of my career.

Over the years, I inevitably found myself on both sides of the pharmaceutical divide. At times I represented big brand-name pharma companies that, with their marvelous patented synthetic or biologic drugs, fought companies that made generics or biosimilars. I helped Big Pharma extend patent lifetimes and enforce them by whatever legal means available. At other times I was on the other side, representing generics or biosimilar companies that were trying to revoke evergreen patents of Big Pharma. If successful, the copiers could enter the market and bring down drug prices earlier than when the patents expired.

Once, I was even fired by the general counsel of a big pharma company when he found out that my law firm also fought on behalf of generic pharma companies—even though he had no fight against any of them. "It's a matter of principle," he said. "We don't work with lawyers who work for generics, *any* generics, even if you may never be our opponent in a case." Thank goodness there are not many biotech and pharma companies on both sides of the divide that are as intolerant as this one—as long as we don't litigate against them, of course.

Drugs such as Enbrel, Humira, Keytruda, or mRNA vaccines against COVID-19, which are frequently the products of academic research, would never see the light of the marketplace without heavy investments from Big Pharma. My work has helped the innovators as well as the developers invest in risky new treatments that have made dramatic improvements in modern medicine. I have also helped remove wrongly issued patents from the marketplace so that prices can come down without waiting for the end of regular patent lives. And more than anything, I have assisted early biotech startups protect lifesaving inventions; without IP, such innovative startups would not even exist.

By lawyering on behalf of these constituencies I came to see problems with the patent system. It is a delicate balancing act. Patents are critical to hedge risk and support investments in research and development. Yet a fine equilibrium needs to be maintained between the innovators' beneficent efforts and their instinct to charge whatever the market will bear for the limited period when they still have patents. If left entirely unregulated, the players in the patent system—especially the big ones—can distort the original premises of limited and controlled exclusive rights.

Happily, during my life as an attorney, I saw major improvements in the US patent system. Submarine patents rarely surface anymore; filings and their prosecutions before the Patent Office are now available online so everyone can see what their competitors are doing and evaluate what they are planning to do; limiting patent lives to twenty years from filing, not seventeen from issuance, has taken the wind out of gaming the system by refiling and refiling, as in the old days.

I believe that with more such thoughtful changes, which maintain a fair balance between the rights of biotech innovators and the rights of patients and consumers, we can make the system work better. This is a complicated picture for a lawyer in private practice. If I wanted to influence political or social decisions in this area I should have been a legislator, not a partner in a law firm.

One groundbreaking—and controversial—legislation is the 2022 enactment of the Inflation Reduction Act (IRA), which includes provisions that, for the first time starting in 2026, will allow the federal government's Medicare and Medicaid to negotiate with manufacturers fair prices for an increasing number of single-source prescription drugs, including patented drugs.[1] Not surprisingly, one of them is our old friend Enbrel, the drug over which I fought at the beginning of my career and which, as I told you in chapter 19, is still protected by a submarine patent that will expire in 2028. Enbrel was chosen by Medicare as one of the first ten drugs for price negotiation.[2] The Enbrel story is far from over.[3]

We live in a time of scarce government funding for pure biological research. Big Pharma have gradually stepped away from their own pursuit of fundamental scientific breakthroughs. Increasingly, they have left that to universities and to small biotech companies. As a result, these entities have become the ultimate source of many biological inventions of great value to society. And intellectual property is one of the central engines for transferring their technology to larger companies and ultimately to the world.

Exploration of biology may no longer be the selfless discipline of old but has, rather, become a blend of the pure and the commercial. As Justice Warren E. Burger said in *Chakrabarty*, the 1980 case dealing with the patentability of genetically engineered live microbes, we cannot stop the scientific mind "from probing into the unknown any more than Canute could command the tides." But 120 years before Burger wrote those words, it was Lincoln who said that the "fire of genius" is not enough for society. It is the patent system that adds the "fuel of interest to the fire of genius." And it's that fuel that will keep the biotechnology engine running.

# Glossary of Scientific Terms

Note: Some of the terms in this glossary are from a database of the US Department of Agriculture; https://www.usda.gov/topics/biotechnology/biotechnology-glossary. Others are from a database and courtesy of the National Human Genome Research Institute (NHGRI); https://www.genome.gov/genetics-glossary/z#glossary. Both databases are in the public domain; both accessed on December 2, 2021.

**5D3 cell line:** A hybridoma cell line that produces monoclonal antibodies of the IgM subclass useful for running hepatitis tests.

**ACGT:** An acronym for the four types of bases found in a DNA molecule: adenine (A), cytosine (C), guanine (G), and thymine (T). A DNA molecule consists of two strands wound around each other, with each strand held together by bonds between the bases. Adenine pairs with thymine, and cytosine pairs with guanine. The sequence of bases in a portion of a DNA molecule, called a gene, carries the instructions needed to assemble a protein.

**Acquired Immunodeficiency Syndrome (AIDS):** A collection of symptoms caused by infection with the human immunodeficiency virus (HIV), which leads to loss of immune cells and leaves individuals susceptible to other infections and the development of certain types of cancers.

**Agribiotech:** A range of tools, including traditional breeding techniques, that alter living organisms, or parts of organisms, to make or modify products; improve plants or animals; or develop microorganisms for specific agricultural uses.

***Agrobacterium tumefaciens:*** The causal agent of gall disease (the formation of tumors) in many species of plants.

**Alfalfa Mosaic Virus, or AlMV:** A worldwide distributed plant virus that can lead to infection and death of a large variety of plant species, including commercially important crops.

**Amino acids:** A set of twenty different molecules used to build proteins. Proteins consist of one or more chains of amino acids called polypeptides. The sequence of the amino acid chain causes the polypeptide to fold into a shape that is biologically active. The amino acid sequences of proteins are encoded in the genes.

**Antibody:** A protein component of the immune system that circulates in the blood, recognizes foreign substances like bacteria and viruses, and neutralizes them. After exposure to a foreign substance, called an antigen, antibodies continue to circulate in the blood, providing protection against future exposures to that antigen.

**Aseptic:** The state of being free from disease-causing microorganisms, such as bacteria, viruses, fungi, and parasites.

**ATCC or American Type Culture Collection:** A nonprofit organization that collects, stores, and distributes standard reference microorganisms, cell lines, and other materials for research and development.

***Bacillus thuringiensis (Bt):*** A soil bacterium that produces toxins that are deadly to some pests. The ability to produce *Bt* toxins has been engineered into some crops.

**Bacteria:** Small single-celled organisms. Bacteria are found almost everywhere on Earth and are vital to the planet's ecosystems. Some species can live under extreme conditions of temperature and pressure ("extremophiles"). The human body is full of bacteria and in fact is estimated to contain more bacterial cells than human cells. Most bacteria in the body are harmless, and some are even helpful. A relatively small number of species cause disease.

**Biologic drug:** Any pharmaceutical drug product manufactured in or extracted from biological sources. Different from totally synthesized pharmaceuticals, biologic drugs include vaccines, gene therapies, tissues, recombinant therapeutic proteins, and living medicines used in cell therapy.

**Biosimilar drug:** A medical product that is almost an identical copy of an original biologic product; it is an officially approved version of the original biologic drug.

***BRCA1/BRCA2:*** The first two genes found to be associated with inherited forms of breast cancer. Both genes normally act as tumor suppressors, meaning that they help regulate cell division. When these genes are rendered inactive due to mutation, uncontrolled cell growth results, leading to breast or ovarian cancer. Women with mutations in either gene have a much higher risk for developing breast cancer than women without mutations in the genes.

**Cancer:** A group of diseases characterized by uncontrolled cell growth. Cancer begins when a single cell mutates, resulting in a breakdown of the normal regulatory controls that keep cell division in check. These mutations can be inherited, caused by errors in DNA replication, or result from exposure to harmful chemicals.

**Cell:** The basic building block of living things. All cells can be sorted into one of two groups: eukaryotes and prokaryotes. A eukaryote has a nucleus and membrane-bound organelles, while a prokaryote does not. Plants and animals are made of numerous eukaryotic cells, while many microbes, such as bacteria, consist of single cells.

**Cellular receptor:** A chemical structure, usually a protein anchored on a cellular membrane, that receives and transmits signals caused by chemical messengers called ligands, which bind to the receptor and cause some form of cellular/tissue response.

**Checkpoint inhibitor:** A cancer drug, usually a monoclonal antibody, that acts to inhibit immune checkpoints, which are key regulators of the immune system. Some cancers can protect themselves from attack by stimulating the immune checkpoints. In contrast, checkpoint inhibitors block the checkpoints, restoring immune system function and allowing the system to attack cancer cells.

**Chimera or chimeric virus:** A virus that contains genetic material derived from two or more distinct viruses. It is a new hybrid microorganism created by joining nucleic acid fragments from two or more different microorganisms in which each of at least two of the fragments contain essential genes necessary for replication.

**Chromosome:** An organized package of DNA found in the nucleus of the cell. Different organisms have different numbers of chromosomes. Humans have twenty-three pairs of chromosomes—twenty-two pairs of numbered chromosomes, called autosomes, and one pair of sex chromosomes, X and Y.

**CRISPR (also known as CRISPR-Cas-9):** A technology that can be used to edit genes within organisms. This editing process has a wide variety of applications including basic biological research, development of biotechnological products, and treatment of diseases.

**Cystic fibrosis:** A hereditary disease characterized by faulty digestion, breathing problems, respiratory infections from mucus buildup, and the loss of salt in sweat. The disease is caused by mutations in a single gene and is inherited as an autosomal recessive trait, meaning that an affected individual inherits two mutated copies of the gene.

**Disarmed *Agrobacterium*:** An *Agrobacterium* that has been genetically engineered to remove the gall cancer-causing genes from the original one.

**DNA, or Deoxyribonucleic Acid:** Chemical name for the molecule that carries genetic instructions in all living things. The DNA molecule consists of two strands that wind around one another to form a shape known as a double helix. The sequence of the bases along the backbones serves as instructions for assembling protein and RNA molecules.

**Double helix:** The description of the structure of a DNA molecule. A DNA molecule consists of two strands that wind around each other like a twisted ladder.

**Enzyme:** A biological catalyst that is almost always a protein. It speeds up the rate of a specific chemical reaction in the cell. The enzyme is not destroyed during the reaction and is used over and over. A cell contains thousands of different types of enzyme molecules, each specific to a particular chemical reaction.

**Erythropoietin, or EPO:** A natural protein that stimulates the formation of red blood cells in the bone marrow.

**Evolution:** The process by which organisms change over time. Mutations produce genetic variation in populations, and the environment interacts with this variation to select those individuals best adapted to their surroundings. The best-adapted individuals leave behind more offspring than less well-adapted individuals.

**Extremophile or extremophilic:** From the Latin, "loving extremes," is an organism able to live in extreme environments, such as high temperatures, radiation, or acidity.

**F1 hybrid:** Also known as filial 1 hybrid, this is a term used in agriculture and is the first filial generation of seed or offspring of distinctly different parental types. F1 hybrids are used in genetics and in selective breeding. Subsequent generations are called F2, F3, etc.

**Fermentation:** In food or beverage production, it is any process using a microorganism such as yeast, in which the activity of the microorganism brings about the production of alcohol to produce beer or wine.

**Fingerprinting using DNA:** A laboratory technique involving DNA, used to establish a link between biological evidence and the identity of a subject. A DNA sample is taken from the evidence at hand and is compared with a DNA sample from a subject. If the two DNA profiles are a match, then the evidence came from that subject. Conversely, if the two DNA profiles do not match, then the evidence cannot have come from the subject. DNA fingerprinting is also used to establish paternity.

**Gall:** An abnormal swelling growth on the external tissues of a plant, sometimes known by the term "crown gall disease," which is similar to benign tumors or warts in animals.

**Gel electrophoresis:** A laboratory technique used to separate DNA, RNA, or protein molecules based on their size and electrical charge. An electric current is used to move molecules to be separated through a gel. Pores in the gel work like a sieve, allowing smaller molecules to move faster than larger molecules.

**Gene:** The basic physical unit of inheritance. Genes are passed from parents to offspring and contain the information needed to specify traits. Genes are arranged, one after another, on structures called chromosomes. A chromosome contains a single, long DNA molecule, only a portion of which corresponds to a single gene. Humans have approximately 20,000–30,000 genes arranged on their chromosomes.

**Gene therapy:** An experimental technique for treating disease by altering the patient's genetic material. Most often, gene therapy works by introducing a healthy copy of a defective gene into the patient's cells.

**Genetic code:** The instructions in a gene that tell the cell how to make a specific protein. A, C, G, and T are the "letters" of the DNA code. Each gene's code combines the four chemicals in various ways to spell out three-letter "words" that specify which amino acid is needed at every step in making a protein.

**Genetic engineering:** The process of using recombinant DNA (rDNA) technology to alter the genetic makeup of an organism. Traditionally, humans have manipulated genomes indirectly by controlling breeding and selecting offspring with desired traits. Genetic engineering involves the direct manipulation of one or more genes. Most often, a gene from another species is added to an organism's genome to give it a desired trait, or phenotype.

**Genetically modified organism (GMO):** An organism produced through genetic modification.

**Genetic modification:** The production of heritable improvements in plants or animals for specific uses, via either genetic engineering or other more traditional methods.

**Genetic testing:** The use of a laboratory test to look for genetic variations associated with a disease. The results of a genetic test can be used to predict or rule out a suspected genetic disease or to determine the likelihood of a person passing on a mutation to their offspring.

**Genome:** The entire set of genetic instructions found in a cell. In humans, the genome consists of twenty-three pairs of chromosomes, found in the nucleus, as well as a small chromosome found in the cell's mitochondria. Each set of twenty-three chromosomes contains approximately 3.1 billion bases of DNA sequence.

**Herbicide-tolerant or -resistant crops:** Crops that have been developed to survive application(s) of particular herbicides by the incorporation of certain gene(s) either through genetic engineering or traditional breeding methods. The genes allow the herbicides to be applied to the crop to provide effective weed control without damaging the crop itself.

**Huntington's disease:** An inherited disease characterized by the progressive loss of brain and muscle function. Symptoms usually begin during middle age. The disease is inherited as an autosomal dominant trait, meaning that a single mutated copy of the responsible gene is sufficient to cause the disease.

**Hybrid seeds:** Seeds of a hybrid plant that has been created by crossing male and female inbred lines.

**Hybrid vigor:** Characteristics of a hybrid plant—such as height, growth, and yield—that its individual parents don't have.

**Hybridoma cell:** A hybrid cell line made by fusing antibody producing B-cells with immortal B-cell cancer cells, a myeloma, the fusion having both the antibody-producing ability of the B-cell and the longevity and reproductivity of the myeloma.

**IgM:** One of several distinct subclasses of antibodies, such as IgG, IgA, IgE, IgD, and IgM.

**Immortalized cell line:** A population of cells isolated from an organism, such as a human being, which would normally die but, due to mutation, can be kept alive and growing indefinitely under laboratory conditions.

**Immune therapy or immunotherapy:** A therapy for cancer that uses the body's own immune system to attack cancer cells.

**Inbred:** Offspring from the mating or breeding of individuals or organisms that are closely related genetically.

**Insecticide resistance:** The development or selection of heritable traits (genes) in an insect population that allows individuals expressing the trait to survive in the presence of levels of an insecticide (biological or chemical control agent) that would otherwise debilitate or kill this species of insect.

The presence of such resistant insects makes the insecticide less useful for managing pest populations.

**Ligand:** A substance that forms a complex with a cellular receptor, like a key binding to its lock, to trigger a biological signal or response.

**LQT Syndrome or LQTS:** A cardiac condition that results in an increased risk of an irregular heartbeat that can result in fainting, drowning, seizures, or sudden death, episodes that can be triggered by exercise or stress.

**Lymphocyte:** A type of white blood cell that is part of the immune system. There are two main types of lymphocytes: B-cells and T-cells. The B-cells produce antibodies that are used to attack invading bacteria, viruses, and toxins.

**Molecular biology:** The study of the structure and function of proteins and nucleic acids in biological systems.

**Monoclonal antibody:** An antibody made by cloning a unique white blood cell. All subsequent antibodies derived this way trace back to a unique parent cell. It is possible to produce monoclonal antibodies that specifically bind to virtually any suitable substance; they can then serve to detect or purify it. In contrast to polyclonal antibodies, which are mixtures of many different antibody molecules, monoclonal antibodies are all chemically identical. This capability has become an important tool in biochemistry, molecular biology, and medicine.

**Monocrop:** The practice of growing the same crop species in the same field year after year.

**Monoculture:** The practice of growing one crop species in a field at a time, although not necessarily the same species year after year, as in mono-cropping.

**Mosaic virus:** Any virus that causes infected plant foliage to have a mottled appearance.

**mRNA or messenger RNA:** A single-stranded RNA molecule that is complementary to one of the DNA strands of a gene. The mRNA is an RNA version of the gene that leaves the cell nucleus and moves to the cytoplasm, where proteins are made. During protein synthesis, an organelle called a ribosome moves along the mRNA, reads its base sequence, and uses the genetic code to translate each three-base triplet, or codon, into its corresponding amino acid.

**Mutation:** A change in a DNA sequence. Mutations can result from DNA copying mistakes made during cell division, exposure to ionizing radiation, exposure to chemicals called mutagens, or infection by viruses.

**Nuclear membrane:** A membrane that encloses the cell nucleus. It serves to separate the chromosomes from the rest of the cell. The nuclear membrane includes an array of small holes or pores that permit the passage of certain materials, such as nucleic acids and proteins, between the nucleus and cytoplasm.

**Nucleic acid:** An important class of macromolecules found in all cells and viruses. The functions of nucleic acids have to do with the storage and expression of genetic information. Deoxyribonucleic acid (DNA) encodes the information the cell needs to make proteins. A related type of nucleic acid, called ribonucleic acid (RNA), comes in different molecular forms that participate in protein synthesis.

**Nucleus:** A membrane-bound organelle that contains the cell's chromosomes. Pores in the nuclear membrane allow for the passage of molecules in and out of the nucleus.

**Organ:** A collection of tissues that structurally form a functional unit specialized to perform a particular function. Your heart, kidneys, and lungs are examples of organs.

**Organic agriculture:** A concept and practice of agricultural production that focuses on production without the use of synthetic inputs and does not allow the use of transgenic organisms. The USDA's National Organic Program has established a set of national standards for certified organic production.

**Paclitaxel or Taxol:** An anticancer drug found in the bark of yew trees.

**PD-1:** (Also known as PD-1 checkpoint) is a cell surface receptor bound on the membrane of immune cells that regulates the immune system's response to other cells of the human body, preventing auto-immune reactions.

**PD-L1:** A ligand that normally binds PD-1 and suppresses the immune system during events such as pregnancy. PD-L1 appears as a membrane-bound molecule on the surface of many cancer cells and when, as such, it binds to PD-1, the cancer cells evade the host immune system.

**Personalized medicine, also known as precision medicine:** A practice of medicine that uses an individual's genetic profile to guide decisions made in regard to the prevention, diagnosis, and treatment of disease. Knowledge of a patient's genetic profile can help doctors select the proper medication or therapy and administer it using the proper dose or regimen.

**PfSPZ:** Acronym for *Plasmodium falciparum* Sporozoites.

***Phaffia rhodozyma*, or simply, *Phaffia*:** A yeast fungus that produces a bright red color so that animals that feed on it—such as salmon, flamingos, shrimp, or lobster—reflect the red-orange pigmentation.

**Plant breeding:** The use of cross-pollination, selection, and certain other techniques involving crossing plants to produce varieties with particular desired characteristics (traits) that can be passed on to future plant generations.

**Plasmid:** A small, often circular DNA molecule found in bacteria and other cells. Plasmids are separate from the bacterial chromosome and replicate independently of it. They generally carry only a small number of genes, notably, some associated with antibiotic resistance. Plasmids may be passed between different bacterial cells.

***Plasmodium falciparum*:** A parasite of humans that causes malaria; it is transmitted through the bite of a female Anopheles mosquito and is responsible for around 50 percent of all malaria cases.

**Promoter:** A sequence of DNA needed to turn a gene on or off. The process of transcription is initiated at the promoter, usually found near the beginning of a gene.

**Proteins:** An important class of molecules found in all living cells. A protein is composed of one or more long chains of amino acids, the sequence of which corresponds to the DNA sequence of the gene that encodes it. Proteins play a variety of roles in the cell, including structural (cytoskeleton), mechanical (muscle), biochemical (enzymes), and cell signaling (hormones).

**Recombinant DNA (rDNA):** A technology that uses enzymes to cut and paste together DNA sequences of interest. The recombined DNA sequences can be placed into vehicles called vectors that ferry the DNA into a suitable host cell, where it can be copied or serve as a blueprint for the production of proteins.

**Restriction enzyme:** An enzyme isolated from bacteria that cuts DNA molecules at specific sequences. The isolation of these enzymes was critical to the development of recombinant DNA (rDNA) technology and genetic engineering.

**Rheumatoid arthritis:** A long-term autoimmune disorder that primarily affects joints. It typically results in warm, swollen, and painful joints.

**Ribonucleic Acid (RNA):** A molecule similar to DNA. Unlike DNA, RNA is single stranded. An RNA strand has a backbone made of alternating sugar (ribose) and phosphate groups. Attached to each sugar is one of four bases—ACGT. Different types of RNA exist in the cell: messenger RNA (mRNA), ribosomal RNA (rRNA), and transfer RNA (tRNA).

**Single point mutation:** A genetic mutation in which a single nucleotide base—such as one of the four genetic alphabet letters, A, T, C, or G—is changed, inserted, or deleted from the sequence of an organism's genome.

**Sporozoites:** In the context of malarial infections, they are cells that develop in the mosquito's salivary glands, leave the mosquito during a blood meal, and enter the human liver, where they multiply.

**Subunit vaccine:** A vaccine that contains non-live purified parts of a pathogenic agent, such as proteins or sugars, necessary and sufficient to elicit a protective immune response. It does not contain the whole pathogen, unlike live attenuated or inactivated vaccine.

**Syndrome:** A collection of recognizable traits or abnormalities that tend to occur together and are associated with a specific disease.

**Synthetics, also known as small molecule drugs:** Drugs that are prepared by chemical synthesis from other compounds.

**Tobacco Mosaic virus, or TMV:** A worldwide distributed plant virus that can lead to infection and death on a large variety of plant species, including commercially important crops, especially tobacco.

**Trait:** A specific characteristic of an organism. Traits can be determined by genes or the environment, or more commonly by interactions between them. The genetic contribution to a trait is called the genotype. The outward expression of the genotype is called the phenotype.

**Transgenic:** Means that one or more DNA sequences from another species have been introduced by artificial means. Animals usually are made transgenic by having a small sequence of foreign DNA injected into a fertilized egg or developing embryo. Transgenic plants can be made by introducing foreign DNA into a variety of different tissues.

**Tumor Necrosis Factor (TNF):** TNF is a natural molecule used by the immune system for cell signaling. If the immune system detects an infection, it causes the release of TNF to alert other system cells as part of an inflammatory response.

**Tumor suppressor gene:** A gene that directs the production of a protein that is part of the system that regulates cell division. The tumor suppressor protein plays a role in keeping cell division in check. When mutated, a tumor suppressor gene is unable to do its job, and as a result uncontrolled cell growth may occur. This may contribute to the development of a cancer.

**Vector:** Any vehicle, often a virus or a plasmid that is used to ferry a desired DNA sequence into a host cell as part of a molecular procedure. Sometimes the term is used to describe an insect such as a mosquito that carries an agent causing a disease such as malaria.

**Viral resistance:** The ability of plants to survive infection by viruses by the incorporation of certain gene(s) into the plants either through genetic engineering or traditional breeding methods.

**Virus:** A small collection of genetic code, either DNA or RNA, surrounded by a protein coat. A virus cannot replicate alone. Viruses must infect cells and use components of the host cell to make copies of themselves. Often, they kill the host cell in the process, and cause damage to the host organism. Viruses cannot be killed by antibiotics; only antiviral medications or vaccines can eliminate or reduce the severity of viral diseases, including AIDS, COVID-19, measles, and smallpox.

# Glossary of Legal Terms

**Amicus brief, also known as an amicus curiae brief:** From the Latin, "friend of the court," this is a legal brief where someone who is not a party to a case assists a court by offering information, expertise, or insight that has a bearing on the issues in the case.

***Argoudelis:*** Legal shorthand for *In re Argoudelis,* a decision from the Court of Customs and Patent Appeals, which in 1970 held that a legally acceptable way to enable a live biological specimen is by an escrow deposit in a cell repository.

**Article of manufacture:** One of the four principal categories of things that may be patented in US patent law. The other three are a process, a machine, and a composition of matter.

**Bayh-Dole Act:** Federal legislation enacted in 1980 that allows universities and other recipients of federal funding to retain ownership of inventions made during the funding period and to grant exclusive licenses to assist with their commercialization.

**Best mode requirement:** In US patent law the disclosure in a patent must include the "best mode" of making or practicing the invention to ensure that the inventor does not retain the best for himself.

**Blocking patent:** A patent relating to an area of technology (e.g., TV) which prevents the patent owner from using an improvement in her own invention (e.g., color TV) because the improvement is covered by a second patent owned by someone else. The commercialization of color TVs is blocked to both.

**CAFC, or Court of Appeals for the Federal Circuit:** The appellate court that has exclusive jurisdiction over cases involving the patent laws of the United States.

***Chakrabarty:*** Legal shorthand for *Diamond v. Chakrabarty*, a decision from the Supreme Court of the United States, which in 1980 held that the aliveness of microbes is no impediment to their patentability.

**Claim:** In a patent, a separate and carefully worded paragraph that succinctly describes the metes and bounds of the legally protected invention.

**Compulsory license:** The right of a government to compel a private maker of a patented drug, against its wishes, to allow another company to make the drug at a lower price.

**Conversion:** A legal wrong that has been made to a person by taking away their property and asserting ownership over it that is inconsistent with the real owner's title.

**Correlation patent:** A patent that is intended to protect the discovery of a correlation between a body indicator such as a gene mutation and disease.

**Cross-licenses:** A simultaneous set of mutual patent licenses given by the owner of one patent to the owner of a second patent and vice versa, when both patents are blocking each other.

**Description requirement:** A requirement of US patent law that compels full description of an invention in exchange for the grant of a patent.

**Designing around:** The legal ability of anyone to redesign a patented article or process so that it avoids infringing the patent.

**Doha Declaration:** A World Trade Organization (WTO) statement that clarifies the scope of the TRIPS Agreement, stating, for example, that TRIPS can and should be interpreted "to promote access to medicines for all."

**Eligibility requirement:** In US patent law an invention needs to be eligible for patenting by falling into certain legal categories—such as a process, a machine, an article of manufacture, or a composition of matter—without falling afoul of three basic exceptions: natural laws, natural phenomena (such as natural products), or abstract ideas.

**Enablement requirement:** In US patent law an invention needs to be described in a patent so that its description will "enable" a person in the field of the invention to be able to readily reproduce it.

**ESCRs:** An acronym that stands for "Economic, Social and Cultural Rights," which arise from an international treaty that came into force in 1976. ESCRs are sometimes known as the "Second Bill of Rights," and include labor rights, rights to social security, to family life, to an adequate standard of living, to health, to free education, and to participation in cultural life.

**Exhaustion or patent exhaustion:** After a product covered by an intellectual property (IP) right, such as by a patent right, has been sold by the IP right owner or by others with the consent of the owner, the IP right is said to be exhausted. It can no longer be exercised by the owner. Exhaustion is also referred to as the first sale doctrine.

**First sale doctrine:** See "Exhaustion or patent exhaustion."

**Genus claim:** A patent claim that protects a family or collection of items able to perform the patented invention. For example, a genus claim in a drug patent may extend to more than one drug and encompass a family or "genus" of similar drugs.

**Injunction:** A court order that compels a party to do, or refrain from doing, specific acts. A party that refuses to comply with an injunction could face monetary sanctions or imprisonment, and can be charged with contempt of court.

**Infringement:** The making, using, or selling of a patented invention without permission from the patent holder, such as a license.

**Interference:** A unique US patent procedure, abolished in 2013, used to determine which among two or more patent applicants on file was the first inventor; after 2013 you simply look at the respective filing dates.

***Myriad Genetics:*** Legal shorthand for *Association for Molecular Pathology* ["AMP"] *v. Myriad Genetics*, a decision from the Supreme Court of the United States, which in 2013 held that isolated genes are not eligible subject matter for patenting.

**Non-obviousness requirement:** In US patent law, a patentable invention needs to be "non-obvious" to a person in the field of the invention so that things that are readily apparent and routinely made from what is otherwise in the public domain also stay in the public domain.

**Novelty requirement:** In patent law a patentable invention must be "novel"; that is, it cannot be published, known, or used anywhere in the world before the earliest filing date of a patent application on it. The only exception is the United States, which gives the inventor a one-year grace period for his or her own acts of disclosure.

**Patent pool:** An arrangement by which at least two companies agree to cross-license patents relating to a particular technology. The creation of a patent pool can save patentees and licensees time and money, and, in case of blocking patents, it may also be the only reasonable method for making the invention available to the public. See also "cross-licenses" and "blocking patent."

**Patent scope:** In US patent law a patent may protect a broad and wide range of different examples of an invention, known as a patent of broad scope; or it may protect only one or two specific examples of the invention, known as a patent of narrow scope.

**Patent thicket:** An overlapping set of patent rights held by one or multiple entities, which requires others that wish to commercialize the invention to reach licensing deals for multiple patents.

**Prior art:** Anything related to an invention that is available to the public, such as by publication, conference, general knowledge or use, video, or lecture, before the filing date of a patent application for the invention.

**Pro bono:** From the Latin "for the public good," it refers to work undertaken voluntarily and without payment to provide legal services for people who are unable to afford them.

**Specific claim:** A patent claim that protects no more than one item able to perform the patented invention. For example, a specific claim in a drug patent extends to no more than one drug and does not extend to other similar examples of such drug.

**Submarine patent:** A patent whose issuance and publication are intentionally delayed by the applicant for a long time. This strategy requires a patent system in which, first, patent applications are not published, and second, patent term is measured from grant date, not from filing date. In the United States, patent applications filed before 1995 were not published and remained secret until they were granted, thus being "submarines" until the day of grant.

**TRIPS Agreement:** Standing for the "Agreement on Trade-Related Aspects of Intellectual Property Rights," it is an international agreement between all the member nations of the WTO, which establishes minimum standards for the regulation of IP. It introduced IP law into the multilateral trading system for the first time.

**Utility or usefulness requirement:** A requirement of US patent law that compels that in exchange for the grant of a patent, an invention must have practical and immediate utility and not be only an object of further research.

***Wands:*** Legal shorthand for *In re Jack R. Wands et al.*, a decision from the Court of Appeals for the Federal Circuit, which in 1988 held that patent claims of a broad scope can be enabled by routine screening.

# Notes

## Introduction

1. Abraham Lincoln, Manner of buoying vessels, U.S. Patent 6,469 (filed March 10, 1849) (issued May 22, 1849). Lincoln's patent claims a mechanism for lifting boats that get stuck on sandbars.

2. Abraham Lincoln, "Lecture on Discoveries and Inventions: Second Lecture on Discoveries and Inventions, February 11, 1859," in *Collected Works of Abraham Lincoln*, ed. Roy P. Basler, vol. 3 (New Brunswick, NJ: Rutgers University Press, 1953), 363, https://www.abrahamlincolnonline.org/lincoln/speeches/discoveries.htm.

3. Diamond, Commissioner of Patents and Trademarks v. Chakrabarty, 447 U.S. 303, 100 S. Ct. 2204 (1980).

## 1. The Invention of Patents

1. Not everyone believes that the Venetian Patent Statute was a fundamental break with prior legal systems. See Joanna Kostylo, "Commentary on the Venetian Statute on Industrial Brevets (1474)," in *Primary Sources on Copyright (1450–1900)*, ed. L. Bently and M. Kretschmer, accessed September 2, 2023, www.copyrighthistory.org.

2. Archivio di Stato (ASV), Senato Terra, reg. 7, c. 32r.

3. U.S. Const., art. I, § 8, cl. 8.

4. An Act to Promote the Progress of Useful Arts, ch. 7, 1 Stat. 109 (April 10, 1790), accessed September 2, 2023, https://govtrackus.s3.amazonaws.com/legislink/pdf/stat/1/STATUTE-1-Pg109.pdf.

5. "Thomas Jefferson to Isaac McPherson, 13 August 1813," in *The Papers of Thomas Jefferson, Retirement Series, March 11 to November 27, 1813*, vol. 6, ed. Jefferson Looney (Princeton, NJ: Princeton University Press, 2009), 379–86, Founders Online, National Archives, https://founders.archives.gov/?q=Jefferson%2C%20Thomas%20Recipient%3A%22McPherson%2C%20Isaac%22&s=1411311111&sa=&r=2.

6. Adam Mossoff, "Who Cares What Thomas Jefferson Thought about Patents? Reevaluating the Patent 'Privilege' in Historical Context," *Cornell Law Review* 92 (July 2007): 953, http://scholarship.law.cornell.edu/clr/vol92/iss5/2.

7. "Paul Berg: Interview." NobelPrize.org, Nobel Prize Outreach AB 2023, accessed March 20, 2023, https://www.nobelprize.org/prizes/chemistry/1980/berg/interview. Reprinted by permission of the Nobel Foundation, © Nobel Web AB 2001.

8. The official name of the agency is the "United States Patent and Trademark Office," but for ease of reading I will refer to it throughout as the "Patent Office."

9. Jason S. Diedering and Saravana B. Kumar of 4C Medical Technologies, Inc., Repositioning wires and methods for repositioning prosthetic heart valve devices within a heart chamber and related systems, devices and methods, US Patent 11,000,000 (filed September 12, 2019) (issued May 11, 2021).

10. The Hon. Randall Rader, Chief Judge, US Court of Appeal for the Federal Circuit, *Keynote Address*, GTIF—Licensing Executives Society International Meeting, Geneva, Switzerland, January 20, 2014.

11. The first time a trademark-registered brand name appears, it is marked as such, "®." Following the first instance, the symbol does not appear again.

12. Chris Woolston, "Generic Drug Savings," *HealthDay*, accessed March 19, 2023, https://consumer.healthday.com/encyclopedia/drug-center-16/misc-drugs-news-218/generic-drug-savings-646392.html.

## 2. The Path from Pure to Commercial

1. Shahal Abbo, Avi Gopher, and Gila Kahila Bar-Gal, *Plant Domestication and the Origins of Agriculture in the Ancient Near East* (Cambridge: Cambridge University Press, 2022).

2. Gregor Mendel, "Versuche über Pflanzenhybriden," *Verhandlungen des naturforschenden Vereines in Brünn* 4 (1865), published in *Abhandlungen* (1866): 3–47.

3. Patrick E. McGovern et al., "Fermented Beverages of Pre- and Proto-historic China," *Proceedings of the National Academy of Sciences of the United States* 101, no. 51 (December 2004): 17593–98, http://doi.org/10.1073/pnas.0407921102.

4. James D. Watson and Francis H. Crick, "Molecular Structure of Nucleic Acids: A Structure for Deoxyribose Nucleic Acid," *Nature* 171, no. 4356 (April 1953): 737–38, https://doi.org/10.1038/171737a0.

5. David A. Jackson, Robert H. Symons, and Paul Berg, "Biochemical Method for Inserting New Genetic Information into DNA of Simian Virus 40: Circular SV40 DNA Molecules Containing Lambda Phage Genes and the Galactose operon of *Escherichia coli*," *Proceedings of the National Academy of Sciences, USA* 69, no. 10 (October 1972): 2904–9.

6. Paul Berg, David Baltimore, Sydney Brenner, Richard O. Roblin III, and Maxine F. Singer, "Summary Statement of the Asilomar Conference on Recombi-

nant DNA Molecules, *Proceedings of the National Academy of Sciences* 72, no. 6 (June 1975): 1981–84.

7. Stanley N. Cohen, Annie C. Y. Chang, Herbert W. Boyer, and Robert B. Helling, "Construction of Biologically Functional Bacterial Plasmids *In Vitro*," *Proceedings of the National Academy of Sciences, USA* 70, no. 11 (November 1973): 3240–44, https://doi.org/10.1073/pnas.70.11.3240.

8. Keiichi Itakura et al., "Expression in *E. coli* of a Chemically Synthesized Gene for the Hormone Somatostatin," *Science* 198 (1977): 1056–62.

9. David D. Goeddel et al., "Expression in *E. coli* of Chemically Synthesized Genes for Human Insulin," *Proceedings of the National Academy of Sciences, USA* 76 (1979): 106–10.

10. "A Missed Opportunity?," *What Is Biotechnology?*, Medical Research Council, accessed March 20, 2023, https://www.whatisbiotechnology.org/index.php/exhibitions/milstein/patents/The-monoclonal-antibody-patent-saga.

11. José E. García-Sánchez, Enrique García, and María L. Merino, *Cien Años de la Bala Mágica del Dr. Ehrlich* (1909–2009)—100 Years of Dr. Ehrlich's Magic Bullet (1909–2009)," *Enfermedades Infecciosas y Microbiología Clínica* 28, no. 8 (October 2010): 521–33, https://doi.org/10.1016/j.eimc.2009.07.009.

12. "Missed Opportunity?"

13. Bradley J. Fikes, "San Diego's Hybritech Still Influences Local Biotech, 40 Years Later," *San Diego Union-Tribune*, May 9, 2018, 6 AM PT.

14. Doogab Yi, *The Recombinant University: Genetic Engineering and the Emergence of Stanford Biotechnology (Synthesis)* (Chicago: University of Chicago Press; Illustrated edition, 2015).

15. City of Hope National Medical Center v. Genentech, Inc., 75 Cal. Rptr. 3d 333 (Supreme Court of California 2008).

16. Mark Vaeck et al., "Transgenic Plants Protected from Insect attack," *Nature* 328 (July 1987): 33–37, https://doi.org/10.1038/328033a0.

17. Jos Bijman, "AgrEvo: From Crop Protection to Crop Production," *AgBioForum* 4, no. 1 (2001): 20-2–5, https://agbioforum.org/wp-content/uploads/2021/02/AgBioForum_4_1_20-1.pdf; and Bayer's corporate website under "history": https://www.bayer.com/en/history/.

18. Peter T. Jones et al., "Replacing the Complementarity-Determining Regions in a Human Antibody with Those from a Mouse," *Nature* 321, no. 6069 (May 29, 1986): 522–25, https://doi.org/10.1038/321522a0.

19. "Biotechnology in the US—Market Size 2006–2029," IBIS World, updated January 26, 2023, https://www.ibisworld.com/industry-statistics/market-size/biotechnology-united-states.

## 3. A Clash of Two Worlds

1. Stanley N. Cohen, Annie C. Y. Chang, Herbert W. Boyer, and Robert B. Helling, "Construction of Biologically Functional Bacterial Plasmids *In Vitro*," *Proc Natl Acad Sci U S A* 70, no.11 (November 1973): 3240–44, https://doi: 10.1073/pnas.70.11.3240.

2. Victor K. McElheney, "Gene Transplants Seen Helping Farmers and Doctors," *New York Times*, May 20, 1974.

3. "Niels Reimers. Stanford's Office of Technology Licensing and the Cohen/Boyer Cloning Patents," an oral history conducted in 1997 by Sally Smith Hughes, PhD, *Regional Oral History Office*, the Bancroft Library, University of California, Berkeley, 1998.

4. Asilomar Conference, *Proceedings of the National Academy of Sciences* 72, no. 6 (June 1975): 1981–84.

5. Stanley N. Cohen and Herbert W. Boyer, Process for producing biologically functional molecular chimeras, U.S. Patent 4237224A, assignee Leland Stanford Junior University (filed November 4, 1974) (issued December 2, 1980) (expired December 2, 1997).

6. "Paul Berg: Interview," *NobelPrize.org*, Nobel Prize Outreach AB 2023, accessed March 20, 2023, https://www.nobelprize.org/prizes/chemistry/1980/berg/interview. Reprinted by permission of the Nobel Foundation, © Nobel Web AB 2001.

7. Paul Berg and Janet E. Mertz, "Personal Reflections on the Origins and Emergence of Recombinant DNA Technology," *Genetics* 184, no. 1 (January 2010): 9–17, https://doi.org/10.1534/genetics.109.112144.

8. Apparently, and notwithstanding his 2010 paper, Paul Berg thought that patenting was fine in some contexts, including gene patents to produce valuable therapeutic proteins. Perhaps Berg was not anti-patent or anti-biotech, but he was indeed worried about commercial incentives intruding on academic research. Personal communication from Professor Robert Cook-Deegan, May 2024.

9. "A 255 Million Patent—Stanford University's Cohen Boyer Patent," *IPVISIONision*, accessed September 2, 2023, https://info.ipvisioninc.com/resources/interesting-patent-maps/cohen-boyer-patent.

10. "A Missed Opportunity?," *What Is Biotechnology?*, Medical Research Council, accessed March 20, 2023, https://www.whatisbiotechnology.org/index.php/exhibitions/milstein/patents/The-monoclonal-antibody-patent-saga.

11. The one-year grace period for one's own publications was confirmed in 1982; see *In re* Katz, 687 F.2d 450 (C.C.P.A. 1982).

12. A. Spinks, Advisory Council for Applied Research and Development (ACARD) / Advisory Board of the Research Councils / Royal Society, "Report of the Joint Working Party" (1980).

13. "Missed Opportunity?"

14. Nicholas Wade, "Inventor of Hybridoma Technology Failed to File for Patent," *Science* 208, no. 4445 (May 16, 1980): 693. https://doi.org/10.1126/science.208.4445.693. Reprinted by permission from AAAS.

15. E. M. Tansey and P. P. Catterall, "Technology Transfer in Britain: The Case of Monoclonal Antibodies—The Transcript of a Witness Seminar Held at the Wellcome Institute for the History of Medicine, London, on 24 September 1993," *Welcome Witnesses to Twentieth Century Medicine*, accessed March 20, 2023, http://www.histmodbiomed.org/witsem/vol1.html.

16. Arti K. Rai and Robert Cook-Deegan, "Racing for Academic Glory and Patents: Lessons from CRISPR," *Science* 358, no. 6365 (November 2017): 874–76. https://doi.org/10.1126/science.aao2468. PMID: 29146800. Reprinted by permission from AAAS.

17. Bayh-Dole Act of 1980, Pub. L. No. 96-517, 94 Stat. 3015 (1980) (codified as amended at 35 U.S.C. § 200–212).

18. For a detailed history of the lobbying for, and enactment of, the Bayh-Dole Act, especially its opponents' arguments, see Alexander Zaitchick, "The Great American Science Heist: How the Bahy-Dole Act Wrested Public Science from the People's Hands," *Intercept*, August 29, 2021, 6 AM, accessed September 2, 2023, https://theintercept.com/2021/08/29/bayh-dole-act-public-science-patents/.

19. "Mission and History," *AUTM*, accessed March 20, 2023, https://autm.net/about-autm/mission-history.

20. Barbara J. Culliton, "The Hoechst Department at Mass. General," *Science* 216 (June 1982): 1200–1203.

21. Culliton, "Hoechst Department," 1200.

## 4. Why Patents in Academia?

1. As noted previously, the rationale that without private investment academic inventions will not benefit the public was central to the enactment of the Bayh-Dole Act; yet not everyone agreed; see Alexander Zaitchik, "American Science Heist": How the Bayh-Dole Act Wrested Public Science from the People's Hands," *Intercept*, August 29, 2021, 6 AM, accessed September 2, 2023, https://theintercept.com/2021/08/29/bayh-dole-act-public-science-patents/.

2. K. Peppel, D. Crawford, and Bruce Beutler, "A Tumor Necrosis Factor (TNF) Receptor-IgG Heavy Chain Chimeric Protein as a Bivalent Antagonist of TNF Activity," *Journal of Experimental Medicine* 174, no. 6 (December 1991): 1483–89, https://doi.org/10.1084/jem.174.6.1483.

3. The $470 million payment came in two tranches: the first, for $186 million in 2006, was for US royalties; see "Amgen Settles Royalty Dispute," *L.A.*

*Times Archives*, June 7, 2006, 12 AM PT, accessed September 3, 2023, https://www.latimes.com/archives/la-xpm-2006-jun-07-fi-amgen7-story.htmland. The second tranche, for $284 million in 2007, was for foreign royalties; see "2011 Index of The Massachusetts Innovation Economy," *Massachusetts Technology Collaborative* (2011): 37, accessed September 3, 2023, https://www.google.com/url?sa=t&rct=j&q=&esrc=s&source=web&cd=&ved=2ahUKEwi84KDW3Y6BAxVMKFkFHWzjAusQFnoECBAQAQ&url=https%3A%2F%2Fstg.masstech.org%2Fsites%2Fdefault%2Ffiles%2F2022-04%2FMAInnovationEconomy_2011.pdf&usg=AOvVaw2enHUB5OkiRqpQfNqOEDVt&opi=89978449.

4. Suzanne Clancy, "Genetic Mutation," *Nature Education* 1, no. 1 (2008): 187, accessed March 22, 2023, https://www.nature.com/scitable/topicpage/genetic-mutation-441/.

5. Robert Sanders, "FDA Approves First Test of CRISPR to Correct Genetic Defect Causing Sickle Cell Disease," *Berkeley News*, March 30, 2021, accessed March 22, 2023, https://news.berkeley.edu/2021/03/30/fda-approves-first-test-of-crispr-to-correct-genetic-defect-causing-sickle-cell-disease/.

6. Walter Isaacson, *The Code Breaker: Jennifer Doudna, Gene Editing, and the Future of the Human Race* (New York: Simon and Schuster, 2021).

## 5. Scientist-Lawyers

1. Mary T. Hannon, "The Patent Bar Gender Gap: Expanding the Eligibility Requirements to Foster Inclusion and Innovation in the US Patent System," *IP Theory* 10, no. 1 (Fall 2020): 1–21, https://www.repository.law.indiana.edu/ipt/vol10/iss1/1.

2. Destiny Peery, "2019 Survey Report on the Promotion and Retention of Women in Law Firms," National Association of Women Lawyers, https://www.nawl.org/page/2018survey.

3. Jill Abramson, "The Princes of the Patent Bar," *American Lawyer* (April 1983): 6–10.

4. Patricia A. Martone, "Reflections on One Woman's Legal Career and the Critical Role of Mentors," *NYIPLA Bulletin* (October/November 2014): 13.

## 6. Who Owns Tangibles Taken from Your Body?

1. John Moore v. The Regents of the University of California, 51 Cal. 3d 120, 271 Cal. Rptr. 146, 793 P.2d 479 (Supreme Court of California 1990).

2. David W. Golde and Shirley G. Quan, Unique T-lymphocyte line and products derived therefrom, U.S. Patent 4,438,032 (filed January 6, 1983) (issued March 20, 1984).

3. Rebecca Skloot, *The Immortal Life of Henrietta Lacks* (New York: Crown Publishing Group, 2011).

4. *The Immortal Life of Henrietta Lacks*, HBO, 2017, https://www.hbo.com/movies/the-immortal-life-of-henrietta-lacks.

5. Catherine Offord, "Henrietta Lacks Estate Sues Thermo Fisher over HeLa Cell Line," *Scientist* (October 2021), accessed December 2021, https://www.the-scientist.com/news-opinion/henrietta-lacks-estate-sues-thermo-fisher-over-hela-69279.

6. See, e.g., "Pierce™ HeLa Protein Digest Standard," Thermo Fisher Scientific, accessed December 2021, https://www.thermofisher.com/order/catalog/product/88329?SID=srch-srp-88329.

7. Meredith Watman, "What Does the Historic Settlement Won by Henrietta Lacks's Family Mean for Others?" *Science Insider*, August 7, 2023, accessed September 3, 2023, https://www.science.org/content/article/what-does-historic-settlement-won-henrietta-lacks-s-family-mean-others.

## 7. Who Owns the Intangibles?

1. Richard Dawkins, *The Selfish Gene* (Oxford: Oxford University Press, 1976).

2. See 35 U.S.C.A. § 102 (b)(1)(A) and (B) (2013).

3. One oft-cited case for this proposition is Jamesbury Corporation v. Worcester Valve Company, 318 F. Supp. 1 (D. Mass. 1970) ("idea reduced to writing after termination of employment held not to be "invention" for purposes of contract," at 7). For a thorough discussion of the ownership of ideas in the context of employment, see Robert P. Merges, "The Law and Economics of Employee Inventions," *Harvard Journal of Law and Technology* 13, no. 1 (Fall 1999): 2–54.

4. I have slightly changed some of the ancillary facts to maintain confidentiality of the individuals involved. However, the central facts of the case are unchanged.

## 8. Quarreling Colleagues

1. Association for Molecular Pathology et al. v. Myriad Genetics, Inc. et al., 569 U.S. 576, 133 S. Ct. 2107 (2013).

2. Wen-Hwa Lee et al., Products and methods for controlling the Suppression of the neoplastic phenotype, U.S. Patent 7,060,688 (filed December 21, 2001) (issued June 13, 2006).

3. Theodore Friedmann and Richard Roblin, "Gene Therapy for Human Genetic Disease?," *Science* 175, no. 4025 (March 1972): 949–55, https://doi.org/10.1126/science.175.4025.949.

4. Petra Kucerova and Monika Cervinkova, *Anticancer Drugs* 27, no. 4 (2016): 269–77. https://doi.org/10.1097/CAD.0000000000000337.

5. *Jim Allison: Breakthrough, Independent Lens* website, PBS, accessed March 25, 2023, https://www.pbs.org/independentlens/documentaries/jim-allison-breakthrough/.

6. Lisa Urquhart, "The Biggest-Selling Drugs of 2023," *Evaluate Vantage*, January 5, 2023, accessed September 4, 2023, https://www.evaluate.com/vantage/articles/analysis/biggest-selling-drugs-2023.

7. The facts and analysis are taken from Dana-Farber Cancer Institute, Inc. v. Ono Pharmaceutical Co., Ltd., Tasuku Honjo, E.R. Squibb & Sons, L.L.C., Bristol-Myers Squibb Company, 964 F.3d 1365 (Fed. Cir. 2020).

8. *Dana-Farber*, 964 F.3d 1365.

9. Mueller Brass Co. v. Reading Industries, Inc., 352 F. Supp. 1357 (E.D. Pa. 1972).

## 9. Patenting Living Things

1. Louis Pasteur, Manufacture of beer and yeast, U.S. Patent 141,072 (filed May 9, 1873) (issued July 22, 1873).

2. Jesse Hirsch, "Goat," *Modern Farmer*, September 16, 2013, accessed September 4, 2023, https://modernfarmer.com/2013/09/saga-spidergoat/.

3. P. J. Federico, "Louis Pasteur's Patents," *Science* 86, 2232 (October 8, 1937): 327.

4. Diamond v. Chakrabarty, 447 U.S. 303, 100 (S. Ct. 2204 1980).

5. *Chakrabarty*, 447 U.S. at 316.

6. *Chakrabarty*, 447 U.S. at 317.

7. In 1985, in *Ex parte* Hibberd et al., 227 U.S.P.Q. 443 (Bd. Pat. App. Int. 1985) the Board of Appeals of the Patent Office held that plants are eligible for patents, and in 1987, in *Ex parte* Allen, 2 U.S.P.Q. 2d. (Bd. Pat App. Int. 1987), the board held that artificial oysters are eligible.

8. *Chakrabarty*, 447 U.S. at 308.

9. Association for Molecular Pathology et al. v. Myriad Genetics, Inc., et al., 569 U.S. 576, 133 S. Ct. 2107 (2013).

## 10. Patenting Genes

1. Association for Molecular Pathology et al. v. Myriad Genetics, Inc. et al., 569 U.S. 576, 133 (S. Ct. 2107 2013).

2. A portion of this chapter was originally published in slightly different form in Jorge A. Goldstein, "Of Isolated Genes and Covalent Bonds: A Personal Memoir of *Myriad Genetics*," *IP Today* 22 (2015): 10–12, www.iptoday.com (magazine and site discontinued).

3. *Myriad Genetics*, 569 U.S. at 576.

4. Marcy E. MacDonald, Christine M. Ambrose, Mabel P. Duyao, and James F. Gusella, Huntingtin DNA, protein and uses thereof, U.S. Patent 5,686,288 (filed May 20, 1994) (issued November 11, 1997).

5. Tsui Lap-Chee, et al., Cystic fibrosis gene, U.S. Patent 6,201,107 (filed June 6, 1995) (issued March 13, 2001).

6. Association for Molecular Pathology et al. v. United States Patent and Trademark Office, et al., 702 F. Supp. 2d 181 (S.D.N.Y. 2010).

7. Association for Molecular Pathology et al. v United States Patent and Trademark Office, et al., 689 F.3d 1303, 1312 (Fed. Cir. 2012).

8. *Association for Molecular Pathology*, 689 F.3d at 1328.

9. Brief for Academics in Law, Medicine, Health Policy and Clinical Genetics as Amici Curiae in Support of Neither Party, *Association for Molecular Pathology* (October 26, 2012).

10. Brief of James D. Watson, PhD, as Amicus Curiae in Support of Neither Party, *Association for Molecular Pathology* (January 31, 2013), at 12.

11. Here is a simplified explanation of cDNA. Genomic DNA contains both introns and exons. Exons are the part of a genomic gene that encode proteins. Introns do not encode proteins. Introns are spliced out during natural transcription from DNA to RNA. RNA is purified in the lab from source material after genomic DNA, proteins, and other cellular components are removed. The purified RNA then serves as a template for cDNA synthesis. cDNA is synthesized through in vitro reverse transcription using an enzyme called reverse transcriptase. As the name indicates, the enzyme takes RNA back to DNA. And since the RNA no longer contains introns, the cDNA made by reverse transcription will be "exons-only." An exons-only cDNA sequence derived from a genomic sequence that originally had exons and introns will be artificial and therefore will be patent eligible. But if the enzyme reverse-transcribes RNA that came from a genomic sequence that was already exons only (e.g., if the RNA template was transcribed naturally from inside an exon), then the cDNA will still be naturally occurring and not eligible.

12. *In Re BRCA1*-and *BRCA2*-Based Hereditary Cancer Test Patent Litigation, University of Utah Research Foundation, et al., v. Ambry Genetics Corporation, 3 F. Supp. 3d 1213 (C.D. Utah 2014), *aff'd*, 774 F. 3d 755 (Fed. Cir. 2014).

## 11. Patenting Pureness

1. Transcript of the April 15, 2013, oral hearing in Association for Molecular Pathology et al. v. Myriad Genetics, Inc., et al., 569 U.S. 576, 133 S. Ct. 2107 (2013), https://dukespace.lib.duke.edu/bitstreams/0be9ada2-dc5c-42b9-94b2-c729fe7c781b/download, *Tr.* at 3. See also Krista Cox, "SCOTUS Oral Arguments in AMP v. Myriad Genetics; Court to Determine Answer to Question: Are Human Genes Patentable?," *Knowledge Ecology International*, April 15, 2013, https://www.keionline.org/22174.

2. Hearing in Association for Molecular Pathology, *Tr.* at 6.

3. Hearing in Association for Molecular Pathology, *Tr.* at 8.

4. Jordan Goodman and Vivien Walsh, *The Story of Taxol: Nature and Politics in the Pursuit of an Anti-Cancer Drug* (Cambridge: Cambridge University Press, 2001).

5. Nicholas J. Talbot, "Plant Immunity: A Little Help from Fungal Friends," *Current Biology Dispatches* 25, no. 22 (2015): R1074–76, https://doi./org/10.1016/j.cub.2015.09.068.

6. Goodman and Wash.

7. "2014 Procedure for Subject Matter Eligibility Analysis of Claims Reciting or Involving Laws of Nature/Natural Principles, Natural Phenomena, and/or Natural Products. Example III. B. Claim 1," March 2014, http://www.uspto.gov/patents/law/exam/myriad-mayo_guidance.pdf.

8. Chenghua Luo and Jorge Goldstein, "Patenting Purified Natural Products by Specific Activity: Eligibility and Enablement," *Bloomberg BNA Life Sciences Law and Industry Report* 9 (2015): 633ff.

9. 2014 Interim Guidance on Patent Subject Matter Eligibility: A Rule by the Patent and Trademark Office, § 3, Nature-based Products, 79 Fed. Reg. 74625–26," December 16, 2014, https://www.federalregister.gov/documents/2014/12/16/2014-29414/2014-interim-guidance-on-patent-subject-matter-eligibility.

10. Natural Alternatives International, Inc. v. Creative Compounds, LLC, 918 F.3d 1338 (Federal Circuit 2019).

## 12. Enabling Life

1. Poppenhusen v. Falke, 19 F. Cas. 1048 (C.C.S.D. N.Y. 1861) (No. 11279).

2. Application of Alexander D. Argoudelis, Clarence De Boer, Thomas E. Eble, and Ross R. Herr, 434 F.2d 1390 (C.C.P.A. 1970).

3. *In re* Jack R. Wands, Vincent R. Zurawski, Jr., and Hubert J.P. Schoemaker, 858 F.2d 731 (Fed. Cir. 1988).

4. Amgen Inc. et al. v. Sanofi et al., 143 S. Ct. 1243 (2023).

5. Judge Newman is in the news as of this writing. At age ninety-six, she is still on the Court of Appeals adjudging cases—or trying to. Several of her fellow judges, however, have claimed that she is no longer able to decide cases properly and have asked her to step down. On May 10, 2023, Newman, fiery dissenter to the end, sued her colleagues on the bench asserting violation of her civil rights. Newman v. Moore, 1:23-cv-01334 (D.D.C. 2023). In September 2023 with the litigation pending, the court suspended Newman from taking on new cases for a year. Judge Newman appealed her suspension to the Judicial Conference of the United States' Committee on Judicial Conduct and Disability. In February 2024 the committee affirmed the suspension. See *In re: Complaint No. 23-90015*, U.S. Judicial Conf. Comm. Rev. Cir. Council Conduct & Disability, February 7, 2024.

6. In early 2024 the Patent Office reaffirmed the validity of using the eight *Wands* factors in enablement analysis and recommended that its examiners continue doing so. See Guidelines for Assessing Enablement in Utility Applications and Patents in View of the Supreme Court Decision in Amgen Inc. et al. v. Sanofi et al., FR Doc. 2024-00259 (January 10, 2024).

## 13. Microbes

1. Gunnard Kenneth Jacobson et al., Astaxanthin over-producing strains of *phaffia rhodozyma*, U.S. Patent 5,466,599 (filed April 19, 1993) (issued November 14, 1995).

2. Ágnes Nagy, Zsuzsanna Palágy, Lajos Ferenczy, and Csaba Vágvölgyi, "Radiation-Induced Chromosomal Rearrangement as an Aid to Analysis of the Genetic Constitution of *Phaffia rhodozyma*," *FEMS Microbiology Letters* 152, no. 2 (July 1997): 249–54. https://doi.org/10.1111/j.1574-6968.1997.tb10435.x.

3. Steven Hedlund, "Joint Venture Formed to Produce Natural Astaxanthin," *SeafoodSource*, February 8, 2009, accessed September 4, 2023, https://www.seafoodsource.com/news/aquaculture/joint-venture-formed-to-produce-natural-astaxanthin.

## 14. Mosquitoes

1. Felgner's team at Vical and his collaborators included Philip Felgner, Jon Wolff, Gary Rhodes, Robert Malone, and Dennis Carson. Ironically, Robert Malone, one of the pioneers of injecting mRNA into animals, has since become a strident anti-vaxxer. See "A Scientist Finds Celebrity in the Anti-Vax Movement," *Washington Post*, A6, January 25, 2022.

2. For a detailed history of the invention of mRNA vaccines from Vical's Dr.

Felgner to BioNTech, Pfizer, and Moderna, see Elie Dolgin, "The Tangled History of MRNA Vaccines," *Nature* 597 (first published September 2021; clarified October 2021): 318–25; or Patricia Thomas, "Naked Came the DNA," chap. 3 in *Big Shot: Passion, Politics, and the Struggle for an AIDS Vaccine* (New York: PublicAffairs Books, 2001), 51.

3. R. Wang, et al., "Induction of Antigen-Specific Cytotoxic T Lymphocytes in Humans by a Malaria DNA Vaccine," *Science* 282, no. 5388 (October 1998): 476–80, https://doi.org/10.1126/science.282.5388.476.

4. See, e.g., Philip L. Felgner, et al., Induction of a protective response in a mammal by injecting a DNA sequence, U.S. Patent 6,214,804 (earliest filing March 21, 1989) (issued April 10, 2001) (expired April 10, 2018). Claim 42 is to a method of generating an antibody response in humans by injecting them with mRNA.

5. "World Malaria Report 2022," Geneva: World Health Organization, 2022, accessed September 4, 2023, https://www.who.int/teams/global-malaria-programme.

6. Andrew Pike, et al., "Changes in the Microbiota Cause Genetically Modified *Anopheles* to Spread in a Population," *Science* 357, no. 6358 (September 2017): 1396–99, https://doi.org/10.1126/science.aak9691.

7. "WHO Takes a Position on Genetically Modified Mosquitoes," World Health Organization, October 20, 2020, accessed March 31, 2023, https://www.who.int/news/item/14-10-2020-who-takes-a-position-on-genetically-modified-mosquitoes.

8. Stephen L. Hoffman, ed., *Malaria Vaccine Development: A Multi Immune Response Approach* (Washington, DC: ASM Press, 1996), 35–75.

9. W. Ripley Ballou, et al., "Safety and Efficacy of a Recombinant DNA Plasmodium Falciparum Sporozoite Vaccine," *Lancet*, June 6, 1987, 1277–81.

10. "WHO Recommends Groundbreaking Malaria Vaccine for Children at Risk," World Health Organization, October 6, 2021, accessed March 31, 2023, https://www.who.int/news/item/06-10-2021-who-recommends-groundbreaking-malaria-vaccine-for-children-at-risk.

11. Michael Finkel, "Stopping a Global Killer," *National Geographic Magazine*, July 2007.

12. Lewis Carrol, *Alice's Adventures in Wonderland*, ed. Edmund R. Brown (Boston International Pocket Library, 1865).

13. Luke Mullins, "Dr. Hoffman vs. the Mosquito," *Washingtonian*, October 23, 2013, https://www.washingtonian.com/2013/10/23/dr-hoffman-vs-the-mosquito.

14. Robert A. Seder, et al., "Protection against Malaria by Intravenous Immunization with a Nonreplicating Sporozoite Vaccine," *Science* 341, no. 6152 (September 2013): 1359–65, https://doi.org/10.1126/science.1241800; Benjamin Mordmüller, et al., "Sterile Protection against Human Malaria by Chemoattenuated Pfspz Vaccine," *Nature* 542, no. 7642 (2017): 445–49, https://doi.org/10.1038/nature21060.

15. Said A. Jongo, et al., "Increase of Dose Associated with Decrease in Protection against Controlled Human Malaria Infection by PfSPZ Vaccine in Tanzanian

Adults," *Clinical Infectious Diseases* 71, no.11 (December 2020): 2849–57, https://doi.org/10.1093/cid/ciz1152.

16. Stephen L. Hoffman and Thomas C. Luke, Apparatuses and methods for the production of haematophagous organisms and parasites suitable for vaccine production, U.S. Patent 8,802,919 (filed March 21, 2007) (issued August 12, 2014).

17. Russell H. Taylor et al., Mosquito salivary gland extraction device and methods of use, U.S. Patent 10,781,419 (filed June 13, 2017) (issued September 22, 2020).

## 15. Plants

1. See, e.g., Kenneth A. Barron, Andrew N. Binns, Mary-Dell M. Chilton, and Antonius J. M. Matzke, Regeneration of plants containing genetically engineered T-DNA, U.S. Patent 6,051,757 (filed January 14, 1983) (issued April 18, 2000). Because of this patent, Chilton was inducted into the National Inventors Hall of Fame in 2015.

2. Anita Rao and Frank Stasio, *The Life, Legacy, and Science of 'Queen of Agrobacterium' Mary-Dell Chilton*, WUNC 91.5 North Carolina Public Radio, February 23, 2015, 11:36 AM EST, accessed September 4, 2023, https://www.wunc.org/show/the-state-of-things/2015-02-23/the-life-legacy-and-science-of-queen-of-agrobacterium-mary-dell-chilton.

3. William E. Timberlake, "Heterosis," in *Brenner's Encyclopedia of Genetics* (2nd ed.), *Science Direct* (2013): 451–53, accessed September 4, 2023, https://doi.org/10.1016/B978-0-12-374984-0.00705-1.

4. Roger N. Beachy, Sue Loesch-Fries, and Nilgun E. Tumer, "Coat Protein-Mediated Resistance against Virus Infection," *Annual Review of Phytopathology* 28 (1990): 451–74.

5. Roger A. C. Jones, "Global Plant Virus Disease Pandemics and Epidemics," *Plants* 10 (2021): 233ff, https://doi.org/10.3390/plants10020233.

6. Nilgun E. Tumer, et al., "Expression of Alfalfa Mosaic Virus Coat Protein Gene Confers Cross-Protection in Transgenic Tobacco and Tomato Plants," *EMBO Journal* 6, no. 5 (1987): 1181–88.

7. L.s. Loesch-Fries, Nancy P. Jarvis, Donald J. Merlo, Gus A.de Zoeten, and John D. Kemp v. Roger N. Beachy, Robert T. Fraley, and Stephen G. Rogers, 104 F.3d 374 (Fed. Cir. 1996).

## 16. Mammals

1. Lewis Carroll, *Through the Looking-Glass, and what Alice Found There* (London: MacMillan and Co., 1882): 75–76.

2. Carlos Alberto Melo, Lino Barañao, Transgenic bovine comprising human

growth hormone in its serum and methods of making, U.S. Patent 7,807,862 (filed September 29, 2004) (issued October 5, 2010).

3. *In re* Roslin Institute (Edinburgh), 750 F.3d 1333 (Fed. Cir. 2014).

4. Off. Gaz. Pat. Off. 24 (April 21, 1987): 1077.

## 17. One Size Does Not Fit All

1. Lisa Larrimore Oullette, "AOC on Pharma and Public Funding," *Written Description* (blog), February 3, 2019, https://writtendescription.blogspot.com/2019/02/aoc-on-pharma-public-funding.html.

2. Oullette.

3. Carl Shapiro, "Navigating the Patent Thicket: Cross Licenses, Patent Pools, and Standard-Setting," in *Innovation Policy and the Economy: I*, ed. Adam B. Jaffe, Josh Lerner, and Scott Stern (Cambridge: MIT Press, 2001): 119–50, *NBER*, https://doi.org/10.1086/ipe.1.25056143.

4. World Intellectual Property Organization, accessed April 3, 2023, www.wipo.int/patents/en/faq_patents.html.

5. "Patent Proliferation: A 30-Year Increase in the Number of Patents per Drug," *OnPoint Analytics*, September 12, 2016, https://onpointanalytics.com/pharma/patent-proliferation.

6. See, e.g., Mayor and City Council of Baltimore et al. v. Abbvie Inc., et al., 42 F.4th 709 (U.S. Court of Appeals, 7th Cir. 2022).

7. Jorge Goldstein, "Biotechnology Patent Pools and Standard Setting," in *Patent Law and Theory: A Handbook of Contemporary Research*, ed. Toshiko Takenaka (Cheltenham, UK: Edward Elgar Publishing Limited, 2009): 712–21; see also Jorge Goldstein, "Patent Pools: Critical Analysis," in *Gene Patents and Collaborative Licensing Models: Patent Pools, Clearinghouses, Open-Source Models and Liability Regimes*, ed. Geertrui van Overwalle (Cambridge: Cambridge University Press, 2009), https://doi.org/10.1017/CBO9780511581182.

8. For a comprehensive discussion of how different industrial constituencies view the role of patents, see *Hearings on "The State of Patent Eligibility in America," before the Subcommittee on Intellectual Property of the US Senate Committee on the Judiciary, Part I*, June 4, 2019, and *Part II*, June 5, 2019, accessed September 5, 2023, https://www.judiciary.senate.gov/committee-activity/hearings/the-state-of-patent-eligibility-in-america-part-i.

9. Misha Angrist, Subhashini Chandrasekharan, Christopher Heaney, and Robert Cook-Deegan, "Impact of Gene Patents and Licensing Practices on Access to Genetic Testing for Long QT Syndrome," *Genetics in Medicine* 12, no. 4 (April 2010 supplement): S111–54, https://doi.org/10.1097/GIM.0b013e3181d68293.

10. Ted J. Ebersole, Marvin C. Guthrie, and Jorge A. Goldstein, "Patent Pools and Standard Setting in Diagnostic Genetics," *Nature Biotechnology* 23, no. 8 (August 2005): 937–38.

11. To provide finer detail to the correlation that was the focus of the dispute between the Mayo Clinic and Prometheus Labs, the relationship was between an individual patient's ability to metabolize a drug and the appropriate dosage of that drug for that patient; at a higher level, this is akin to the effectiveness of that drug for that patient.

12. "Witnesses before House IP Subcommittee Call for Research Exception for Gene Patents," *Bureau of National Affairs, Medical Research Law and Report*, vol. 6, November 7, 2007, 21.

13. "Witnesses before House IP Subcommittee," 21ff.

14. Angrist S121.

15. Mayo Collaborative Services v. Prometheus Laboratories, Inc., 566 U.S. 66, 132 S. Ct. 1289 (2012).

## 18. Controlled Monopolies

1. Aaron S. Kesselheim and Jerry Avorn, "Letting the Government Negotiate Drug Prices Won't Hurt Innovation," *Washington Post*, September 22, 2021.

2. In the Case of NORVIR® Manufactured by Abbott Laboratories, Inc., National Institutes of Health Office of the Director, July 2, 2004. The NIH director in 2004 was Elias Zerhouni. In 2012 the NIH received a second request for march-in for the same drug. It denied it again. See [In] the Case of NORVIR® Manufactured by AbbVie, National Institutes of Health Office of the Director, November 1, 2013. The NIH director in 2012 was Francis Collins.

3. "Letter from Elizabeth Warren, US Senator, Angus S. King, Jr, US Senator, and Lloyd Doggett, Member of Congress, to Gina Raimondo, Secretary, Department of Commerce and Xavier Becerra, Secretary, Department of Health and Human Services," June 9, 2023, accessed September 5, 2023, https://www.warren.senate.gov/oversight/letters/senators-warren-king-and-representative-doggett-seek-answers-from-hhs-and-commerce-on-interagency-working-group-for-bayh-dole.

4. Megan Van Etten, "Setting the Record Straight on the Bayh-Dole Act and March-In," *PhRMA* (blog), May 10, 2023, accessed September 5, 2023, https://phrma.org/en/Blog/Setting-the-record-straight-on-the-Bayh-Dole-Act-and-March-in.

5. *SingleCare* website, accessed January 28, 2023, https://www.singlecare.com/prescription/celecoxib.

6. *SingleCare* website, accessed January 28, 2023, https://www.singlecare.com/prescription/celecoxib.

7. Farah Stockman, “Our Drug Supply Is Sick. How Can We Fix It?” *New York Times*, September 18, 2021.

8. *Drugs.com* website, accessed January 28, 2023, https://www.drugs.com/availability/generic-celebrex.html.

9. Teva Pharmaceuticals International GmbH v. Eli Lilly and Company, 8 F.4th 1349 (Fed. Cir. 2021); Eli Lilly and Company v. Teva Pharmaceuticals International GmbH, 8 F.4th 1441 (Fed. Cir. 2021).

## 19. The Age of Biologics

1. “Scientific Discussion for Approval of Enbrel,” updated October 1, 2004. European Medicines Agency, updated October 1, 2024, originally accessed March 23, 2023, https://www.ema.europa.eu.

2. Bruce A. Beutler, Karsten Peppel, and David F. Crawford, DNA encoding a chimeric polypeptide comprising the extracellular domain of TNF receptor fused to IgG, vectors, and host cells, U.S. Patent 5,447,851 (filed April 2, 1992) (issued September 5, 1995) (expired September 5, 2012).

3. Manfred Brockhaus, et al., Human TNF Receptor Fusion Protein, U.S. Patent 8,063,182 B1 (originally filed in Switzerland on September 12, 1989, having a latest US filing date on May 19, 1995, issued on November 22, 2011, and expiring on November 22, 2028).

4. Charlotte Harrison, “Enbrel Patent Surfaces,” *Nature Biotechnology* 30, no. 2 (February 2012): 123.

5. Immunex Corporation, Amgen Manufacturing, Limited, Plaintiffs-Appellees. Hoffmann-La Roche Inc., Plaintiff v. Sandoz Inc., Sandoz International GmbH, Sandoz GmbH, 964 F.3d 1049 (Fed. Cir. 2020).

6. Jonathan Gardner, “A Three-Decade Monopoly: How Amgen Built a Patent Thicket around Its Top-Selling Drug,” *BioPharma Dive*, November 1, 2021, https://www.biopharmadive.com/news/amgen-enbrel-patent-thicket-monopoly-biosimilar/609042/.

7. Mayor and City Council of Baltimore et al. v. Abbvie Inc., et al., 42 F.4th 709 (U.S. Court of Appeals, 7th Cir. 2022).

8. See, e.g., FTC v. Actavis, Inc., 570 U.S. 136 (2013).

9. The [Indian] Patents Act, 1970, ch. II, “Inventions Not Patentable,” § 3(d), “What Are Not Inventions,” accessed September 5, 2023, https://ipindia.gov.in/writereaddata/Portal/ev/sections/ps3.html, states: “The mere discovery of a new form of a known substance which does not result in the enhancement of the known efficacy of that substance or the mere discovery of any new property or new use for a known substance or of the mere use of a known process, machine or apparatus

unless such known process results in a new product or employs at least one new reactant."

10. Novartis AG v. Union of India, (2007) 4 MADRAS L.J. 1153, http://www.scribd.com/doc/456550/High-Court-order-Novartis-Union-of-India.

## 20. International Access to Patented Biologics

1. Laurence R. Helfer and Graeme W. Austin, "The Human Right to Health, Access to Patented Medicines, and the Restructuring of Global Innovation Policy," ch. 2 in *Human Rights and Intellectual Property: Mapping the Global Interface* (Cambridge: Cambridge University Press, 2011), 90–169.

2. "Data, Life Expectancy at Birth, Male (Years)," World Bank, accessed June 2021, https://data.worldbank.org/indicator/SP.DYN.LE00.MA.IN?view=map&year=2019.

3. Pharmaceutical Manufacturers Association of South Africa v. The President of South Africa et al., Case no. 4183/98 In the High Court of South Africa (Transvaal Provincial Division, 1998), www.cptech.org/ip/health/sa/pharmasuit.html.

4. Mandisa Mbali, "TAC in the History of Rights-Based, Patient Driven HIV/AIDS Activism in South Africa," in *Passages* (Ann Arbor: MPublishing, University of Michigan Library, June 2005), accessed September 5, 2023, https://quod.lib.umich.edu/p/passages/4761530.0010.011/--tac-in-the-history-of-rights-based-patient-driven-hivaids?rgn=main;view=fulltext passages.

5. Lisa Forman, "'Rights' and Wrongs: What Utility for the Right to Health in Reforming Trade Rules on Medicines?" *Health and Human Rights* 10, no. 2 (December 2008): 37–45, https://www.hhrjournal.org/2013/09/rights-and-wrongs-what-utility-for-the-right-to-health-in-reforming-trade-rules-on-medicines.

6. *Agreement on Trade-Related Aspects of Intellectual Property Rights as Amended by the 2005 Protocol Amending the TRIPS Agreement, Art. 31, Sections (a) and (h): Other Use Without Authorization of the Right Holder*, WTO doc., accessed April 9, 2023, https://www.wto.org/english/docs_e/legal_e/trips_e.htm#art3.

7. Jay Taylor, "Compulsory Licensing: A Misused and Abused International Trade Law," *PhRMA* (blog), May 16, 2017, accessed September 6, 2023, https://phrma.org/Blog/compulsory-licensing-a-misused-and-abused-international-trade-law.

8. Ellen Fm't Hoen, Jacquelyn Veraldi, Brigit Toebes, and Hans V. Hogerzeil, "Medicine Procurement and the Use of Flexibilities in the Agreement on Trade-Related Aspects of Intellectual Property Rights, 2001–2016," *Bulletin of the World Health Organization* 96, no. 3 (March 2018): 185–93, https://doi.org/10.2471/BLT.17.199364.

9. Ann Danaiya Usher, "South Africa and India Push for COVID-19 Patents

Ban," *Lancet* 396, no. 10265 (December 5, 2020): 1790–91, https://doi.org/https://doi.org/10.1016/S0140-6736(20)32581-2.

10. Rachel Silverman, "Waiving Patents on Coronavirus Vaccines Won't Help Inoculate Poorer Nations," *Washington Post,* March 21, 2021; Christopher Rowland, Emily Rauhala, and Miriam Berger, "Drug Companies Defend Vaccine Monopolies in Face of Global Outcry," *Post,* March 20, 2021; Joseph E. Stiglitz and Lori Wallach, "Preserving Intellectual Property Barriers to Covid-19 Vaccines Is Morally Wrong and Foolish," *Post,* April 25, 2021.

11. Jorge Goldstein, "Waiving Covid-19 Vaccine Patents Won't Solve the Global Need," *Bloomberg Law,* August 17, 2021, https://news.bloomberglaw.com/ip-law/waiving-covid-19-vaccine-patents-wont-solve-the-global-need.

12. Danaiya Usher, 1790–91.

13. Adam Taylor and Claire Parker, "U.S. Drugmaker Merck to Share License for Experimental Covid-19 Treatment with Nonprofit Organization," *Washington Post,* updated October 27, 2021, 10:54 AM EDT.

14. Mia Rabson, "From Science to Syringe: COVID-19 Vaccines Are Miracles of Science and Supply Chains," *Canadian Press,* February 27, 2021.

15. Peter J. Hotez and Maria Elena Bottazzi, "A COVID Vaccine for All," *Scientific American Online,* December 30, 2021, https://www.scientificamerican.com/article/a-covid-vaccine-for-all.

16. Joe Palca, "Whatever Happened to the New No-Patent COVID Vaccine Touted as a Global Game Changer?" *All Things Considered,* NPR WAMU 88.5, August 31, 2022, 3:36 PM ET, accessed September 6, 2023, https://www.npr.org/sections/goatsandsoda/2022/08/31/1119947342/whatever-happened-to-the-new-no-patent-covid-vaccine-touted-as-a-global-game-cha.

17. Joe Palca, "A Texas Team Comes Up with a COVID Vaccine That Could Be a Global Game Changer," *All Things Considered,* NPR WAMU 88.5, January 5, 2022, 12:47 PM ET, accessed September 6, 2023, https://www.npr.org/sections/goatsandsoda/2022/01/05/1070046189/a-texas-team-comes-up-with-a-covid-vaccine-that-could-be-a-global-game-changer.

18. Eleni Smitham and Amanda Glassman, "The Next Pandemic Could Come Soon and Be Deadlier," August 25, 2021, Center for Global Development, accessed April 5, 2023, https://www.cgdev.org/blog/the-next-pandemic-could-come-soon-and-be-deadlier; Ian Sample, "Bill Gates Call [*sic*] for Huge Global Effort to Prepare for Future Pandemics," *Guardian,* November 3, 2021, https://www.theguardian.com/us-news/2021/nov/04/bill-gates-call-for-huge-global-effort-to-prepare-for-future-pandemics.

19. Jonathan J. Darrow, Michael S. Sinha, and Aaron S. Kesselheim, "When Markets Fail: Patents and Infectious Disease Products," *Food and Drug Law Journal* 73, no. 3 (2018): 361.

20. Scott Berinato, "Moderna v. Pfizer: What the Patent Infringement Suit Means for Biotech," *Harvard Business Review Online*, September 16, 2022, accessed September 6, 2023, https://hbr.org/2022/09/moderna-v-pfizer-what-the-patent-infringement-suit-means-for-biotech.

21. Alan B. Bennett, "Reservation of Rights for Humanitarian Uses," ch. 2.1 in *Intellectual Property Management in Health and Agricultural Innovation* (Oxford: MIHR: 47–62, at 47.

22. Bennett, 52.

23. "Gavi Approach Creates Tiered Pricing for Vaccines," GAVI, the Vaccine Alliance, accessed April 5, 2023, https://www.gavi.org/news/media-room/gavi-approach-creates-tiered-pricing-vaccines.

## 21. Genetically Modified Crops

1. There is some controversy on the origin of the *Phytophtora* culprit, with recent work suggesting that it came from central Mexico. See Amanda C. Saville, Michael D. Martin, and Jean B. Ristaino, "Historic Late Blight Outbreaks Caused by a Widespread Dominant Lineage of *Phytophthora infestans* (Mont.) de Bary," *PLOS ONE* 11, no. 12 (December 28, 2016): e0168381, https://doi.org/10.1371/journal.pone.0168381.

2. Charles C. Mann, *1493: Uncovering the New World Columbus Created* (New York: Alfred A. Knopf, 2011), 221–26.

3. Amalia Leguizamón, "Roundup Ready Nation: The Political Ecology of Genetically Modified Soy in Argentina" (PhD diss., City University of New York, 2014); see also Ryan Hanrahan, "Brazil, Argentina Crop Production Estimates Cut," Farm Policy News, February 23, 2024, accessed April 19, 2024, https://farmpolicynews.illinois.edu/2024/02/brazil-argentina-crop-production-estimates-cut/.

4. The analogy between extractive resources—like oil, gas, or minerals—and agricultural resources—like soybean crops—is imperfect, in that the first are nonrenewable and the second are renewable. Nevertheless, several commentators have analogized soybean crops as part of a "resource curse" that includes both extractives and renewables. See, e.g., Melissa Mittelman, "The Resource Curse," *Bloomberg Online*, September 12, 2014, 9:02 AM EDT, updated May 19, 2017, 2:47 PM, accessed September 6, 2023, https://www.bloomberg.com/view/quicktake/resource-curse#xj4y7vzkg. I use the comparison as a way to argue that when a country is dependent on the price that other countries pay for its extractive or renewable commodities, its economy becomes distorted.

5. Carey Gillam, *Whitewash: The Story of a Weed Killer, Cancer, and the Corruption of Science* (Washington, DC: Island Press, 2017).

6. Vernon H. Bowman v. Monsanto Co., 569 U.S. 278, 133 S. Ct. 1761 (2013).

7. *Bowman v. Monsanto*, at 1763.

8. Organic Seed Growers and Trade Association, et al. v. Monsanto Company and Monsanto Technology LLC, 718 F.3d. 1350 (Fed. Cir. 2013).

9. This assertion is correct insofar as the first-generation seed (F1) can be replanted and that it grows true to form. If the GM crop is a hybrid, as explained in chapter 15, the F1 seed does not grow true to form, and the farmer, regardless of patents, needs to buy another year's worth of seed.

10. Raquel L. Chan, Julieta V. Cabello, and Jorge I. Giacomelli, HAHB11 provides improved plant yield and tolerance to abiotic stress, U.S. Patent 10,738,318 B2 (January 8, 2018) (issued August 11, 2020).

11. Fernanda G. Gonzalez et al., "Field-Grown Transgenic Wheat Expressing the Sunflower Gene Hahb4 Significantly Outyields the Wild Type," *Journal of Experimental Botany* 70, no. 5 (2019): 1669–81, https://doi.org/10.1093/jxb/erz037.

## 22. *Tikkun Olam*

1. David Shatz, Chaim I. Waxman, and Nathan J. Diament, eds., *Tikkun Olam: Social Responsibility in Jewish Thought and Law* (Lanham, MD: Jason Aronson, 1997).

2. "State of the Union Message to Congress," January 11, 1944, Franklin D. Roosevelt Library, accessed September 7, 2023, http://www.fdrlibrary.marist.edu/archives/address_text.html.

3. International Covenant on Economic, Social and Cultural Rights, January 3, 1976, Part III, Art. 15.1., 993 U.N.T.S. 14531, 9.

4. This portion of the chapter was originally published in slightly different form in Jorge A. Goldstein, "Protecting Rainforest-Derived Technology Equitably," *WIPO Magazine* (February 2019): 32–39. Reprinted with permission from WIPO under CC BY 3.0 IGO. Attribution 3.0 IGO.

5. Ulrike Hellerer and K. S. Jarayaman, "Greens Persuade Europe to Revoke Patent on Neem Tree. . . ," *Nature* 405 (May 18, 2000): 266–67.

6. Convention on Biological Diversity, December 29, 1993, Arts. 2 and 3, 1760 U.N.T.S. 30619, 146–47.

7. For a clear statement of opposition to the Convention on Biological Diversity, see, e.g., Laura Reifschneider and Roger J. Marzulla, "The Biodiversity Treaty Challenges Intellectual Property Rights," *Federalist Society*, July 1, 1997, accessed September 7, 2023, https://fedsoc.org/commentary/publications/the-biodiversity-treaty-challenges-intellectual-property-rights.

8. "Public Interest Intellectual Property Advisors (PIIPA)," United Nations—Department of Economic and Social Affairs, accessed April 7, 2023, https://sdgs.un.org/partnerships/public-interest-intellectual-property-advisors-piipa.

## 23. Foliar Feeding

1. "Letter from Walter Plecker (December 1943)," *Encyclopedia Virginia*, accessed April 20, 2024, https://encyclopediavirginia.org/primary-documents/letter-from-walter-a-plecker-to-local-registrars-et-al-december-1943/.

2. Walter Plecker, 1943.

3. E. F. Schumacher, *Small Is Beautiful: Economics as If People Mattered* (New York: Perennial Library / Harper and Row, January 1, 1973).

4. David Hosansky, "Intermediate Technology," *Encyclopaedia Britannica* Online, updated June 2, 2014, https://www.britannica.com/technology/intermediate-technology.

## Epilogue

1. Inflation Reduction Act, Pub. L. No. 117-169, 136 Stat. 1833f (August 16, 2022).

2. "Fact Sheet: Biden-Harris Administration Announces First Ten Drugs Selected for Medicare Price Negotiation," The White House, August 29, 2023, accessed September 7, 2023, https://www.whitehouse.gov/briefing-room/statements-releases/2023/08/29/fact-sheet-biden-harris-administration-announces-first-ten-drugs-selected-for-medicare-price-negotiation/.

3. As of this writing, AstraZeneca, Merck, and nine other pharmaceutical companies have each filed lawsuits over the IRA against the US government. See "Inflation Reduction Act," O'Neill Institute, *Health Care Litigation Tracker*, accessed April 22, 2024, https://litigationtracker.law.georgetown.edu/issues/inflation-reduction-act. The companies argue, among other things, that it is improper to call something a "negotiation" when there is a threat of penalties for disagreeing. For a response to the pharma companies' litigation complaints, see Ross Daval, C. Joseph, and Aaron S. Kesselheim, "We Can't Let Drug Companies Get Out of Negotiating Prices," *Washington Post*, October 23, 2023, 5:45 AM EDT. On March 1, 2024, the United States District Court for the District of Delaware dismissed AstraZeneca's lawsuit for lack of standing. More important, the court found that AstraZeneca could not demonstrate that the IRA deprived it of a constitutionally protected property interest. AstraZeneca Pharmaceuticals LP et al. v. Becerra et al., Memorandum Opinion, Case 1:23-cv-00931-CFC (US District Court for the District of Delaware, filed March 1, 2004), accessed April 22, 2024, https://litigationtracker.law.georgetown.edu/litigation/astrazeneca-pharmaceuticals-lp-v-becerra-et-al-2/.

# Index

Note: Figures are indicated by page numbers in *italics*.

# About the Author

Jorge Goldstein is a patent attorney trained in molecular biology who began his career on the ground floor of the biotech revolution forty years ago. He is a partner in Sterne, Kessler, Goldstein, and Fox, PLLC, and the author of the casebook *US Biotechnology Patent Law*, now going into its ninth edition from Thomson Reuters. He earned a PhD in chemistry from Harvard University and a JD from George Washington University Law School.